CRACKING NATURE'S CODE

The Potential Answer to Everything

Josephine Head

JB Head PhD

Balboa Press books may be ordered through booksellers or by contacting:

Balboa Press
A Division of Hay House
1663 Liberty Drive
Bloomington, IN 47403
www.balboapress.co.uk
1 (877) 407-4847

Print information available on the last page.

ISBN: 978-1-9822-8064-2 (sc)
ISBN: 978-1-9822-8065-9 (e)

Balboa Press rev. date: 05/22/2019

Thank you to all of those who said NO or that have been instrumental otherwise. It's because of them I'm doing it myself.

If

If you can keep your head when all about you
Are losing theirs and blaming it on you;
If you can trust yourself when all men doubt you,
But make allowance for their doubting too:
If you can wait and not be tired by waiting,
Or, being lied about, don't deal in lies,
Or being hated don't give way to hating,
And yet don't look too good, nor talk too wise;

If you can dream- -and not make dreams your master;
If you can think- -and not make thoughts your aim,
If you can meet with Triumph and Disaster
And treat those two impostors just the same:.
If you can bear to hear the truth you've spoken
Twisted by knaves to make a trap for fools,
Or watch the things you gave your life to, broken,
And stoop and build' 'em up with worn-out tools;
If you can make one heap of all your winnings
And risk it on one turn of pitch-and-toss,
And lose, and start again at your beginnings,
And never breathe a word about your loss:
If you can force your heart and nerve and sinew
To serve your turn long after they are gone,
And so hold on when there is nothing in you
Except the Will which says to them: 'Hold on!'

If you can talk with crowds and keep your virtue,
Or walk with Kings- -nor lose the common touch,
If neither foes nor loving friends can hurt you,
If all men count with you, but none too much:
If you can fill the unforgiving minute
With sixty seconds' worth of distance run,
Yours is the Earth and everything that's in it,
And- -which is more- -you'll be a Man, my son!

Rudyard Kipling

Contents

State of Play: Exciting By Design in Theory- Not Much Fun in Practice!

Baby Grow Up- Life is Designed to be Exciting!

In the evolutionary scheme of things, humans are immature babies. We have so far missed the point of our personal and collective purpose as responsible managers of the planet through our disconnection with the laws of nature. We have found ourselves in a scary, unsafe and unsustainable space, on a human race to the bottom. Hanging onto the current story of life by our fingertips, we have become so busy doing we have so little opportunity to being human; it seems impossible to imagine that life was designed to be exciting and there could be so much more to it. Time to Wake Up and Grow Up!

Chaos Reigns Across the Board- Reboot Required.

We have surrounded ourselves with climate change, food shortages, ensuing mass migration, conspiracy theories, cover ups, political nightmares, pointless jobs and general non-sense. Human culture, analogous to a bacterial culture, has enjoyed periods of exponential growth, which if not enriched, becomes toxic and dies. Historically we have recovered to exponential growth through conflict and world wars. We cannot afford to sit and wait for history to repeat itself. We can tweak the system and delay world war for as long as possible. Releasing the full potential of People Power to reboot the system for a common sense cultural redesign and live life to its full really exciting potential, as nature designed, is truly vital now. IF! We are willing to cut the umbilical cord and let go of capitalism.

Human Story Hijack and I'm alright Jack to High 5!

The human evolutionary story has been hijacked by the concept of 'survival of the fittest', an ensuing DNA bandwagon and the invention of money, amongst other prime suspects- well past its expiry date. Accelerated by the 'evolutionary self-correction' of current politics, the world appears to be waking up to the flaws and opportunities of a cultural redesign fit for the future by getting back on track with the inherent design laws of nature to create an upward spiral to success. If only we could achieve the full infinite potential of our human capital – People Power built on natural highs - and optimise finite global resources with a clear way ahead. What a wonderful world it could be.

Order from Chaos- Potentially Messy Transition.

The need for and development of a strategy for a non-catastrophic transition plan to a new human story of success and get back on evolutionary track with nature was identified in 'Changing Images of Man', the elements of which still hold good today (Markley and Harman, 1982). One of many very unfortunate examples that we have all the answers in unburied treasure, but no current means of making them openly visible and transparent to build on. My ideas have the potential to change that.

The document highlights the limitations of classical science being broken down in a reductionist framework as being not fit for purpose when applied to highly complex systems, as well as failure to recognise consciousness and spirituality in order to fully realise human potential in a unified approach. It identifies that new physical principles need to be discovered to breakdown separate frontiers of science, with the computer as an extension of the human nervous system, which in turn will allow augmentation of human intellect as being key to success- **Check Mate!**

Cracking Nature's Code: In a Nutshell

The actin cytoskeleton provides the centre-piece of the puzzle missing as the Universal architect, the physical form for information of life at its origin, evolution as 'selection of the successful', consciousness, being human and quantum mechanics. Information is conducted from the environment to the cell through the membrane, co-operating with the immune system and mitochondria at the front end, orchestrating a physical process of information triangulation (What If!) and instruction of DNA at the back-end, in a dynamic energetic and physical feedback loop for continuous improvement in order to optimise physiological adaptation to the environment. In essence it provides your unique sense of self and provides inner guidance to success, self-satisfaction and a sense of purpose.

A scientific synthesis crystallises the science pieces I have connected to support an embryonic unifying theory of the nature of life as information and our material reality. The incredibly finely tuned cell architecture made up of actin is the missing piece of the puzzle making sense of a multiplicity of mysteries.

What If! Actin as the Universal Architect cannot be experimentally proven to be the case - it doesn't really matter. Mastering the mysteries is key to success: Switching focus from the genetic blueprint should be fruitful in prevention and treatment of disease, disorder, distress and depression. Its unique properties are starting to be realised with novel applications being developed to allow us to switch from depleting our finite natural resources and potential solutions to current crises not yet imagined.

Proof of Principle for Biomimicry- Simple Solution to Solve Complexity

At the very least the theory / hypothesis is robust enough to provide a proof of evolutionary principles of the simple design rules governing the human operating system and a case study of what stretching your imagination can do. If me myself and I can get this far by **wondering What If!** Triangulating information and stretching my imagination, it seems impossible to imagine the potential of harnessing our collective ability and how different life might be. Hopefully it will become evident in hindsight.

Biomimicry of the process into a technological wisdom-based 'success framework' as a thinking tool to complement and enhance our natural creativity and imagination, to find the sweet spot between human and artificial intelligence, to catalyse the full collective force of 'People Power' potential is what I have in mind. **A simple solution to solve complexity is on the cards.**

A single 'success' framework as a standard operating procedure with a continuous improvement feedback loop embedded in a 3 layered model to reflect the micro/macro/mega-evolution levels of complexity provides inter-operability as a foundation for cultural redesign. Trust, transparency and reward through Blockchain has massive potential power to create a viable social safety.net with the internet, to get your just reward for all your effort and realise Universal Basic Assets- much more imaginative and realistic than Universal Basic Income proposed to fill the widening economy gaps that Artificial Intelligence is creating.

How beautiful to my mind that biomimicry of the universal process and the root cause of life that caused the Cambrian explosion has the potential to become the root cause of a successful Cambrian explosion in AI and the transformation of human kind to human-unity with a simple solution to potentially solve the complexity of everything. Emergence in technology has a real potential to create a peaceful (R)EVOLutionary Epochalyptic leap and avoid the potential Artificial Intelligence apocalypse we are otherwise in danger of doing and becoming victims of our own success.

People Power Reboot Solution to the Rescue- The Potential Answer to Everything. The solution of engaging People Power to reboot globalisation has loosely been identified in principle by the World Economic Forum and United Nations. As far as I am aware, there are no pragmatic cards on the table other than mine. Built on nature's simple design principles of life as information by Cracking Nature's Code, biomimicry in technology has the real potential to facilitate and accelerate the transition as quickly and painlessly as possible to catalyse cultural change in an upward spiral to a new human story of success: The Potential Answer to Everything.

Time to Get Excited! Nature designed life that way. So much potential to look forward to: Learning to light up your natural curiosity, find your sense of purpose, happiness and sense of satisfaction. Create exciting ideas in a new social medium, generating natural highs, so you can become centre of attention for all the right reasons and make a real difference. No more meaningless jobs. Make life more deep and meaningful, caring and sharing. Facilitate and accelerate breakthroughs and keep the planet safe for the future.

Dystopian Gaps are Widening- Protopian Potential. Dystopian scientific predictions for 2020 are for increasing intra-elite competition, elite overproduction and political polarization. We are already seeing increased social instability through the stagnation and decline of living standards and the decline of fiscal health of the state (Turchin, 2017). The need is even more urgent to create a new story. Not unsuccessful utopian compliant human constructs such as Marxism and communism, but one of a protopian future. We can make life as great as possible as quickly as is humanly possible by realising full human potential.

Labour of LOVE. This book reflects an extraordinary sequence of highly unexpected events in my own EVOLutionary life path and something of a culture shock to my system. My wakeup call and labour of LOVE started in 2015 by chance. The sequence of sections should take you on a journey of understanding and imagination as to the reasons why this is potentially a really good idea to attract the right attention to progress and we can all live happily ever after.

Dear Millennial and Everyone That Cares. Get Connected, Tuned-In and Turned On: Life is designed to be exciting! Following nature's design laws, exciting should become the new normal. We just need to remove the energy blocks created by the bitter-sweet artificial sweetener of money by human design to create the freedom to be your unique self- the ultimate luxury.

The solution I have designed has the potential to exercise your natural curiosity, stretch your imagination to find your sense of self, passion, create novel ideas, connect with other like minds for extra excitement and be rewarded for your effort and value of contribution.

We can change the game from Me, Me, Me, to being meaningful, happier and save the planet along the way. We have the opportunity to get out of the maze we have created to be truly Amazing! Something I hope you realise is worth the effort of putting your heart and soul into.

It's Not Rocket Science. To achieve our full purpose and action potential we must apply full force in the right direction of common sense of collective purpose – it's not rocket science that holds the answer- we do! We just need to learn how to harness and drive to manage the transition by breaking through current barriers and fulfil our role as the wise managers of the evolutionary process and the planet we are designed to be.

Time to Get the Party Started!

People Power Delivering Protopian Potential- (R)EVOLutionary

What If! Imagine the Future is Wonderful and Exciting

Let's re-imagine a new story of success what would it look like? How about discovering the wonder of you! Realising the secrets to health and happiness, where life is defined by purpose and motivation, where we could capture creativity and imagination, through more playtime – nature's grand design- where contribution and best effort are valued and prized. A case for optimism and a cause for excitement not fear.

Cultivating curiosity quotient (CQ) as well as emotional quotient(EQ), not just IQ (Intelligence Quotient), in order to develop your hungry mind the ultimate tool to produce simple solutions for complex problems (Chamorro-Premuzic, 2014), to find your passion (Howes, 2018) and the ultimate freedom of time to play and have fun!

Thought processes culminating in major breakthroughs have historically been very slow to materialise as a function of unconscious creative thinking processes of the individual. A case for thinking without consciousness (Dijksterhuis and Strick, 2016) provides fascinating insights as to the underlying principles as to how this system operates: Alignment with success goal pursuit as the objective; playing around with spatial organisation and polarisation of thought into far-reaching concepts, beyond the normal stretch of day to day conscious thinking; reward for trains of thought that are complex, important or interesting to the individual. This wholly reflects the process by which I naturally stretched my own knowledge and imagination to derive ideas. Through triangulation and extrapolation of existing information – **wondering What If!** to advance science through imagination (Bongiorni, 2017) which turned out in my case to fill some big empty spaces.

The exciting prospect is that we have the opportunity to harness this extra-ordinary human capacity into more rapid conscious thought through technology, to facilitate and accelerate breakthroughs by expanding the collective space far, far, far and away beyond the current limited 'adjacent possible' of current innovation patterns (Loreto et al., 2016).

Magic Potential!

Following nature's design laws, exciting should become the new normal. We just need to remove the energy blocks created by the bitter-sweet artificial sweetener of money by human design to create the freedom to go there: The Ultimate Luxury.

Using just your imagination or technology to facilitate, for triangulation of information and **wondering What If!** to think about the impossible, connecting the apparently unconnected, you can create an embryonic idea/ intuitive hunch. When you get lucky, the biological process makes the electrical connection, completes the circuit and your reward system gives you a natural high- AHA! –nature's nudge to give you the confidence to take personal action or share and expand your ideas with like minds - a luxury I have not been able to enjoy- Ah! Synchronicity is nature's secret way of achieving successful action potential. Sharing ideas in the innovation process of collective purpose creates bigger highs and synchronicities. When you get really lucky you are in the realms of a breakthrough, worthy of escalation to global innovation platforms to make a real difference and the potential for positive contagious spread.

The magical power of light-bulb AHA! moments and synchronicity is something I have learned to experience, the wonders of which make going back to a mundane job impossible, and the desire to share the potential excitement and the motivation. Coincidentally it is having an open mind and observation of the seemingly mundane that has historically revealed secrets of the subconscious inner genius to the conscious mind in creating previously unimaginable breakthroughs. How exciting! Paying attention to your subconscious could be the new normal human operating system.

Protopian Cultural Redesign on the Cards

Getting back on track with nature's inherent design by effectively providing a convergence of design and systems thinking is vital to co-create a cultural redesign towards Protopia (Shermer, 2018) – a new story of successful planetary culture (Seibert, 2018) avoiding the pitfalls of Utopian human conformist constructs.

Unlocking human intelligence 'People Power' potential as citizen scientists (Felt, 2016; National Academies of Sciences, Engineering, and Medicine, 2018) in conjunction with AI has the potential for scaling up across cultural domains, to be the biggest innovation (Johnson, 2016) as a catalyst of cultural change, whilst avoiding the possibility of opening Pandora's box by taking control of AI's technological potential self-direction that we are otherwise in danger of doing and avoid becoming the ghost in the machine (Sætra, 2018).

(R)EVOLutionary Science Thinking

This is vital to changing our current trajectory to catastrophe and the fortunes of planet Earth in order to get back on track with nature's inherent design. The structure of scientific revolutions (https://en.wikipedia.org/wiki/The_Structure_of_Scientific_Revolutions) shows a clear path to what I think we are looking for to change the rules of the Game. Not a hard painful slog through theory followed by data collection and a quick tweak. Cracks and anomalies start to appear and some bright spark comes up with something much more interesting, satisfying and exciting. Revolutions are often initiated by an outsider- someone not locked into the current model, which hampers vision. Such a quantum shift cannot realistically account for everything but is enough to attract attention and cause others to jump ship. (Unfortunately usually has to go through rocks been thrown at it!) With an embryonic but sufficiently robust hypothesis towards a Unifying theory of Universal connectivity with the origin of life, driver of evolution, consciousness and intelligence provides proof of principle for the inherent design process, puts me in the same (rocky) boat. Recognising the root cause has the potential for monumental breakthroughs in science and medicine and the potential to cure all manner of cultural diseases and crises by getting re-connected with the laws of nature.

(R)EVOLutionary Solution

Design thinking for creatively solving problems to find innovative solutions was (R)EVOLutionary when it was introduced in the 1960's, but has outgrown its usefulness and will require systems thinking to deliver solutions to the complex changes we are now facing. From 30 years of personal experience of building and re-designing business systems to optimise efficiency and inter-operability for success, I have adopted parallel thinking into the design process.

The science and the development of the solution co-evolved through systematisation. Biomimicry of the inherent design process of emergent self-organisation, embedded in technology provides an idea for a solution to facilitate and accelerate the transition with minimum pain. A standardised approach is vital for success, to facilitate people participation, inter-operability to optimise value and transparency to remove barriers to innovation and the cultural immune reaction to novel ideas slowing down the process, as well as providing an inherent mechanism for rewarding effort and value of contribution- delivering the potential for realising Universal Basic Assets, much more imaginative and realistic than Universal Basic Income (UBI), for when you (fortunately!) have no job to go to. We desperately need a peaceful (R)EVOLutionary solution to avoid a messy transition and the potential for World War in the nick of time.

Play to Human Advantage

Despite the negative press of an Artificial Intelligence Apocalypse (Dowd, 2017), or feeling useless without a job to go to (Bregman, 2017), life could be so much sweeter. Playing to the strengths of AI and accentuating our human advantage, people empowerment (Keywell, 2017), to create a complementary sweet spot - where AI does the boring hard work, validates a human hunch, evaluates novelty and impact of human creativity and we get to exercise our imaginations and do the interesting, exciting and valuable stuff, to find, follow and be rewarded for our passions, optimise human compassion, with more time to play.

If you prefer to stay (safe?) stuck in a job, this is worth a read (Houser, 2018). Taking full human advantage of abstract thinking and curiosity, we can potentially reap the valuable rewards of AI (Nanterme, 2017). Breaking down current barriers to science, innovation and change delayed by the current cultural immune reaction if your imaginative ideas are too far-fetched will facilitate and accelerate breakthroughs to improve quality of life, health and avoid a global apocalypse.

Ahead of the Cultural Exponential Growth Curve: Caught in an Evolutionary trap

The connection between the quantum electromagnetic field, biology and consciousness is exciting yet controversial, intimidating in the absence of a safe, anonymous space to explore ideas, before sharing with like minds, and potential reward for effort and value, beyond self-funded self-satisfaction, as would be the case with the technology I have designed.

Collaboration is critical to success. If me, myself and I can get this far in getting to the edges of current science knowledge and beyond, just imagine what connecting People Power can achieve in regenerating an exponential cultural growth curve.

According to Utopia for Realists (Bregman, 2017) which makes a great case for the rationale and viability of a new utopian system and how to get there in principle, (focussing on Universal Basic Income as the solution), it is the duty of thinkers to offer alternative success models and this is the critical moment to make progress. Thinkers are required to have the courage to be utopian. Ideas, however outrageous, have changed the world and they will again. If we want to change the world we need to be unrealistic, unreasonable and impossible. Yesterday's crazy (e.g. slavery, same-sex marriage) has become today's common sense.

I am a big thinker and these are my ideas. Better to kick a ball around than no ball at all on an empty field and a sea of blank faces spectating wondering what life is supposed to be about.

I am very hopeful that broadcasting my message will prevent it falling foul, into the evolutionary trap of money and lack of imagination that frequently befalls innovation – how lucky information theory didn't fall foul! It was an incredibly near miss (Aftab et al., 2001). It's hard to imagine life without it as the source of electromagnetic wave communication tools that we all take for granted.

Time to Wake Up and Grow Up!

Humanity at a Crossroads: Left to Chance or Right by Design?

Baby Grow Up! Mother Nature Knows Best

In the evolutionary scheme of things, humans are immature babies. We have been here a very short time and caused lots of damage. Buckminster Fuller got it right by design some time ago. We have learnt by trial and error how the system is supposed to operate and have yet to get a proper grip, let alone stand up, find our feet, hold our own and be held accountable as responsible wise managers of the planet. We have so far missed the point of our personal and collective purpose through our disconnection with the Universal forces of nature and have found ourselves in a scary, unsafe and unsustainable space, on a human race to the bottom. He identified the design principles of geodesic domes, physical 'tensegrity' and triangulation- **What If!** Turning knowledge into wisdom.

Understanding the nature of the inherent design of our biology to achieve our full potential and embed it into a technology creative thinking tool is my solution to achieving our purpose and planetary success- an Operating Manual for Spaceship Earth, first published in 1969 (Fuller, 2017). I have unwittingly followed in his footsteps and like him (Fuller, 2004; first published 1983) have been my own guinea-pig. He predicted that one person's breakthrough would pave the way for 99,999 others- more than enough to cover the unemployment gaps coming from the advent of Artificial Intelligence.

> "We should do away with the absolutely specious notion that everybody has to earn a living. It is a fact today that one in ten thousand of us can make a technological breakthrough capable of supporting all the rest. The youth of today are absolutely right in recognizing this nonsense of earning a living. We keep inventing jobs because of this false idea that everybody has to be employed at some kind of drudgery because; according to Malthusian Darwinian theory he must justify his right to exist. So we have inspectors of inspectors and people making instruments for inspectors to inspect inspectors. The true business of people should be to go back to school and think about whatever it was they were thinking about before somebody came along and told them they had to earn a living."- Buckminster Fuller

Premature- Now is the Right Time

I think his ideas were premature. I propose a better understanding of the nature of the inherent design of life as information to achieve our full potential. However embryonic it is robust enough for biomimicry in technology as a 'multi-purpose People Power toolkit', to facilitate and accelerate what might otherwise be a messy transition. We have new technology toys to play with, offering much greater potential: With the invention of the internet, advent of AI and the untapped potential of Block chain, the time is right by design to create a shared future of safe and secure Human-unity with an internet Safety. Net in a fractured world, not the futility we have created.

Mother Nature's Magic! Simple Design Recipe for Life as Information

Forget about your smart phone, the most sophisticated technology you've never seen is buried in nature. Understanding how it operates is critical to changing the fatally flawed human operating system (Johnson, 2018). The science story of genetic inheritance we have been sold is falling apart and it turns out we still know so little about how nature operates. The Magic! Of Life as an Information Framework as the secret ingredient, combining physics with biology is just starting to catch on (Davies, 2019). It is still holding on to the genetic story, not knowing the answer to the really tough question as to how life's hardware can write its own software. I had no idea as to how far, far, away we are from the truth in our understanding before I started searching for answers puzzling me and stumbled across so many science puzzles. The simple answer potentially lies with the same smart biological material operating according to nature's simple design principles. (The heavyweight of supporting scientific evidence connecting the physics and biology pieces of the puzzle of life as information, with actin as the universal architect is provided in the section: Cracking Nature's Code: An Embryonic Potential Physiological Theory of Everything. Designed for those already in the best know how.)

Genetic DNA Dilemma- Genotype to Phenotype

Genetics is not all it's cracked up to be. Genetics alone is likely to be barking up the wrong tree to investigate cellular evolution (Liebeskind et al., 2017) or to create a true reflection of a universal evolutionary 'Tree of Life'. More like a genetic bush than a tree (Rokas and Carroll, 2006).

Genetics is struggling to explain the uniqueness of being human (O'Bleness et al., 2012), let alone explaining the uniqueness of being yourself. Scientists have been jumping off the DNA bandwagon as the driver of evolution for some time. The human genome project set out to provide the potential answer to everything health wise and was not nearly as successful as it was supposed to be. It turns out that the complexity of humans is not attributable to our genes and we are as complex as a microscopic worm has come as a bit of a surprise (Barras, 2016).

Evidence now suggests that the era of the 'selfish gene' is now over and has been replaced with the concept of an orchestrating organism within a self-organising non-linear physical process (Latham, 2017). The pieces of information I have collated takes this to another level. Evolution by natural selection turns out to be more 'selection of the successful' through environmental adaptation rather than 'survival of the fittest' through random genetic mutation. Phenotype (outward appearance) through epigenetics (non-genetic influence) is more clearly becoming the right direction.

Back to Basic Cell Biology- Magic!

If you have any idea of what the inside of one of your cells looks like, it is probably made up of an elastic outer membrane, rather like a balloon, which contains a variety of organelles (baby organs) suspended in cytoplasm physically separate from each other. Essentially they have different jobs performing different functions producing different ingredients to create a Magic! recipe for life.

The nucleus contains your genetic blueprint, your DNA (Deoxyribonucleic Acid) information hardware that you inherited from your parents which forms the molecular basis of life as is generally understood. Watching this real time animation might make you wonder HOW? DNA knows what information to process and HOW? The process is synchronised to make sure all the ingredients make their way to be in the right space at the right time https://www.youtube.com/watch?v=fpHaxzroYxg. Biology is only just waking up to their being a secret ingredient to life that provides the mechanism for the organelles to communicate with each other as a dynamic dance, forcing a dramatic rethink and exciting future in cell biology (Dolgin, 2019). The original discovery was made 30 years ago and met with scientific scepticism. Lucky for me I knew about it 30+ years ago secreted in my Ph.D thesis (Head, 1987) and have only just switched on to the possibilities (Surprise! My Dummy Run Section).

Another key organelle is the mitochondrion, the energy powerhouse, which also contains DNA and comes from your mother (sperm only contribute DNA). They have an interesting part in the evolutionary story, having originally being free living single cell bacteria (Prokaryotes) that were subject to evolutionary natural selection and engulfed into a mutually beneficial (symbiotic) relationship to form multi-celled organisms (Eukaryotes) https://www.khanacademy.org/science/biology/structure-of-a-cell/tour-of-organelles/v/mitochondria-video. Bacteria use a process called quorum sensing as a communication system, effectively a decision making process as to whether to take action. This allows them to successfully cooperate with each other and select a favourable growth environment, giving mitochondria another very important role, being actively involved in cell communication as a cost/risk benefit mechanism and efficient use of energy (Is it worth the effort?; Chandel, 2015).

Cytoplasmic Secret Cell Architecture- Space Time Management

I would be surprised if you have any idea of the amazing architecture within the cell cytoplasm https://www.khanacademy.org/science/biology/structure-of-a-cell/tour-of-organelles/v/cytoskeletons

The idea of physical forces being responsible for cellular form and function under the influence of the laws of physics (mechanobiology) was proposed 100 years ago. This was partly ignored due to lack of available technology to investigate and recently gained scientific interest with the realisation that mechanical forces from the environment on the membrane have (at least) as much influence as genetics (Eisenstein, 2017). Without a physical architecture for space time management, how else would the cell, let alone the body manage to get the right ingredients to the right space/place at the right time?

Magic! Of Cell Division.

Taking off DNA's histone protein coat that otherwise stops it working and unwound the double helix and stretched the DNA in one cell all the way out, it would be about 6 feet/2m long. All the DNA in all your cells put together would be about twice the diameter of the Solar System. An amazing feat of engineering is required.

This animation demonstrates the process with cytoskeleton microtubule dynamics centre stage https://www.youtube.com/watch?v=X_tYrnv_o6A. The answers to the unknowns are simple when you know the identity of the secret ingredient. It decides when the time to divide is right and synchronises the process.

Secrets of Quantum Physics and the Secret of the Secret - Scientific Jury Out on Natures Laws- Game Changing Potential

A host of biological mysteries are thought to be attributable to the secrets of quantum physics https://www.youtube.com/watch?v=q4ONRJ1kTdA. Magnetic fields are known to be responsible in bird migration but nobody knows How? The potential answer to that example and others in the film clip is much easier when you think you know the secret ingredient.

The foundation of the 'Orch OR' theory of quantum consciousness rests on microtubules (Hameroff and Penrose, 2014), which describes actin as a gel. To be totally fair, the theory was published in the mid 1990's, way before science started to wake up about actin's potential (Galkin et al., 2012). The Mind to Matter book (Church, 2018) gives chapter and (Uni)verse on the science and symptoms of our amazing creative potential, including the Secret to the Secret, the Law of Attraction and what a game- changer it could be if it was harnessed. Microtubules are touched on as the root cause in the absence of a more sensible explanation. The science jury is still out on nature's laws (Brooks, 2015).

Actin the Magic! Secret Ingredient- The Universal Architect of Life as Information

Joining the disconnected pieces of science scattered across so many different separate disciplines, following my own curiosity without constraints has allowed me to get a glimpse of the biggest universal connectivity picture: **Actin was naturally selected at the origin of life for its information processing capability, as the root cause of universal consciousness, driver of evolution through natural selection, being human and connection with quantum mechanics and universal forces.**

IT Factor Potential- Don't Blame your Genius on Genetic Birth Right!

Hereditability of actin networks as an evolutionary self-organisation process (Maly and Borisy, 2001), hereditability of collective behaviour in harvester ants (Gordon, 2016) and epigenetic inheritance of stress (Lacal and Ventura, 2018) provide strong evidence for actin as the missing piece of the epigenetic evolutionary puzzle. Forget IQ- The real trick is to find our own unique IT factor potential and make it worth the effort. I have one up my sleeve.

Actin the Original Smart Material

The physics of actin gives fascinating insight as to how smart it is. As well as being a dynamic tension sensor, it has lots of unique attributes: Synchronisation of activities, magnetic properties of attraction and resistance and a great candidate as an information processor by virtue of its fast phase transition from wave to particle tread milling dynamics. Its explosive nature has likely been instrumental in driving evolution through energy barriers such as the transition from single cells to multiple celled organisms, Cambrian explosion and the development of evolutionary innovation of the human brain.

Evolutionary success of a species by virtue of actins space and time management, explosive powers and synchronisation to achieve full action potential is something you might be aware of without even realising it. Otherwise how do baby turtles know that the time is right to hatch to ensure there are enough of them to successfully get past predators? Otherwise how do insects successfully communicate, in order for ants to create bridges with their bodies? Or bees communicate with flowers to efficiently collect nectar and distribute pollen for their mutual success, without running out of buzz? Otherwise how do flowers know when is the right time to burst into bloom? Otherwise how do birds know when is the right time to collapse their wings to squeeze through tight spaces or keep so beautifully equally spaced and synchronised timing in flock formation and murmuration?

Actin is the most abundant protein in most eukaryotic cells (Dominguez and Holmes, 2011) and has been largely overlooked, principally down to its unassuming size and limitations in available scientific techniques. The protein atlas provides a clearer picture in human cells (https://www.proteinatlas.org/humancell/actin+filaments).

Actin has only relatively recently been identified in bacteria, creating an artificial evolutionary knowledge gap (De Souza, 2012). There has also been hot debate surrounding whether it stretched as far as the DNA in the nucleus, only recently resolved (Galkin et al., 2012). This has precluded it from having a potentially universal integrated informational role that I propose.

This animation gives a better flavour of actins finely tuned wiring and dynamic nature compared to the much bigger, stronger, less flexible girders of the microtubules https://www.youtube.com/watch?v=tO-W8mvBa78. Actin development supports that of microtubules (Colin et al., 2016) furnishing simple design principles underlying the control of signalling pathways through self-organization to tune signalling processes, so it should not be too surprising that actin potentially provides the real answer to the root cause of consciousness and other biological quantum effects. A common evolutionary link between actin, tubulin and histone proteins (Gardiner et al., 2008) also supports the idea that actin came first, giving rise to tubulins and DNA's histone coat in the evolutionary process.

Science to Drive Intentional Change- Right by Design- Not Just An Un/Happy Accident

My idea has a beautiful fit with that of a 'Science for Intentional Change' by combining design and systems thinking to evolve the future in order to achieve long-term sustainable growth in the global economy and the welfare of the global population (Wilson et al., 2014). Its development has been delayed by the lack of a conceptually unifying theoretical framework of phenotypic plasticity that enables organisms to respond adaptively to their environments. What If! That is what I have managed to do.

Information 'Success' Framework of Phenotypic Plasticity - Adaptive Action by Simple Design

The pieces of the science puzzle I have connected indicate DNA is a blueprint (information hardware). It operates under instruction/ construction to manufacture proteins on demand through inter-play with actin (information software- the physical form for information- you're unique self IT Factor) in a dynamic feedback loop. This optimises energy efficient environmental adaptation to the environment, with the change proportionate to the environmental pressure and the ability to revert back to a more successful design.

[The heavyweight of supporting science is in Cracking Nature's Code, designed more for the benefit of those in the science know how.] Only recently a beautiful example comes from embryology where the cells design is highly dynamic, too complex to be understood by genetics. This usefully breaks down current understanding with some big surprises of plasticity, with dynamic feedback loops rather than a genetically programmed cell outcome (Cepelewicz, 2018a). Much more to do with electromagnetic guidance by the 'foam-like' cell architecture (Lenne and Trivedi, 2018) I propose. Indeed actin has now been shown to be the driving force in mouse embryos, presenting exciting new ways to think about the emergence of cell fates during mammalian development (Frum and Ralston, 2018). Otherwise how do cells know when is the right time to start and stop dividing and what shape and form they are supposed to be? Otherwise how does nature manage to create such beautiful symmetrical design? Otherwise how and why does an embryo retain valuable evolutionary memory (we have gill like structures and tails as part of our embryological journey) to make sure it doesn't repeat previously less successful designs?

A smooth and energy efficient continuous improvement cycle makes a lot more sense to me than a random genetic mutation as the root cause of natural selection and adaptation and forms a robust foundation for the success framework I have designed.

Nature's Nudges- Your Self IT Factor

Nature is giving us nudges in the right direction, to be true to your unique self IT factor. Ignored at our peril and exciting when in tune and buzzing with our inherent sense of purpose and feeling valuable. Rather find that than the disease of Selfitis.

Left Feeling Deflated Below Form

Failure to learn and grow, or repeating patterns of bad behaviour will be punished, with the penalty proportionate to the crime. Better understanding of the fundamental biology will give us the tools to cure all sorts of diseases, both physical and psychological, by addressing root cause. Applying sticking plasters on symptoms by popping pills and artificial stimulants is not the answer to getting us back on track with nature- the fundamental flaw. Not much Fun! Depression is a prime example: Psychologists think they have found its purpose: It has recently been proposed that it is an adaptation for analysing complex problems, as a mechanism to establish inner purpose and restore equilibrium and harmony. The scientific opinion provides an evolutionary systems theory that the mechanism is to avoid nasty surprises. **What If!** Actin as a tension sensor potentially provides the root cause as an internal pressure calibration system setting up an expectation of the experience- pleasantly surprising when experience exceeds expectation and disappointing when it doesn't. Surprising or not, that our immune pattern recognition system designed to protect us from an adverse environment is turning on itself. It appears to be operating as our internal police mechanism through pattern recognition to keep us moving in the right direction.

Right On – Take the Highway- The Sky's the Limit!

We have an inbuilt reward system for getting it right by inherent design. Neurons communicate with the immune pattern recognition system using a type of 'adrenaline rush'. Brain wave pattern harmonisation explains the feel good factor of tuning in to psych-acoustic medicine, achieving relaxation and stress release, increased learning and creativity, memory, and other desirable benefits (otherwise achieved by taking LSD) to achieve a heightened state of consciousness. The pineal gland is a special case: Like your eyes, it has the visual architecture of rods and cones to process light despite being smack bang in the middle of our brain. If connected, this triangulation (a universal evolutionary triangulation architecture for information processing networks) turns it on. The patterns visualised under the influence of psychedelic drugs appear to support my idea- an amazing side-effect of paying attention. It receives the most blood flow of any organ in the body, apart from the kidneys, is one of the largest producers of serotonin but receives very little scientific attention.

Relax- Stretching the Imagination is Exhausting

From experience I know how energetically expensive and exhausting throwing out the actin network is to stretch your imagination, but sufficiently satisfying to make it feel worth the effort. Luckily the only pressure imposed on me at the moment is self-created, so I can take a break, relax my actin architecture and start again.

Conspiracy theories suggest the current system is a set up to keep us all busy doing to stop us busy being human. I can safely say there is no fear of that so the powers that be can relax on the matter. Conspiracies may be real or imagined as an excuse for failure. I really hope that understanding the simplicity of nature's design and a complementary cultural redesign takes away the excuses and allows us to realise our full collective potential.

Science Practice- Not Heading In the Right Direction

Synthetic biology is taking a cell-free approach to speed up innovation (Lu, 2017) - Really!

The human connectome project has painstakingly been identifying the brains connections to test ideas about how brain structure relates to function. It started in 2009 and the leading scientist got his wish of $40million dollars in funding- I wish! Intelligence is just being understood as a function of the gaps in brain connectivity (Williams, 2017). Time to throw out the focus on pulling at its wires and re-think the gaps that actin appears to fill.

Should scientific proof of my embryonic theory/ hypothesis really be necessary as has been suggested by practising Professors, new investigative tools, including 3-D cell culture (Zujur et al., 2017), nanoscale tweezers (Nadappuram et al., 2018), tissue transparency (Eisenstein, 2018), actin engineering (Kumar and Mansson, 2017), 3D bioelectronics (Pitsalidis et al.,2018) and predictive machine learning applied to stem cells (Maxmen, 2017) are now available to harness natures design, develop artificial photosynthesis, prevent and cure disease and distress and develop novel applications - perfect timing.

Natural complexity and interconnectivity is such that we are outgrowing the traditional science method and we need to resist the temptation of testing everything to death in a slow, painful, expensive, inefficient and often non-productive process. Open up to data driven methods, greater rewards for evidence synthesis and hypothesis based approaches according to the weight of evidence wherever that evidence may come from- would open the door for ideas like mine!

Secret of My Success- Mindmapping Mental Dexterity

The secret of my success was using a mindmapping tool- MindGenius- without which this would have been mission impossible. I find it extremely easy to use and has allowed me to intuitively capture thoughts and information inputs from a whole host of disparate sources, make connections, re-organise information and visualise otherwise invisible patterns, coincidences and analogies and create lines of evidence - the mental Dexterity to go Right by Design! Not to under-estimate or substitute for the hard work, pain and perseverance on the journey. New features include the ability for multiple user interaction and an impact/effort assessment to facilitate action plans from your ideas generated, promising to provide an even better fitness for purpose.

We have the ingredients; we just need to change the recipe for success.

Cracking Nature's Code: An Embryonic Potential Physiological Theory of Everything

I Wonder What If! Actin is the Universal Architect of Life as Information

Evidence provided suggests that actin is the universal 'smart' material before DNA popped up. Various theories have been presented as to where DNA came from, with viruses as a strong candidate (Forterre et al., 2000-2013). The science behind quantum biology and consciousness is highly contested and widely considered to be the greatest challenge for modern science. The most widely accepted theory of the nature of how information works is according to the integrated information theory, in the absence of a physical mechanical underpinning process catalysed by energy. Cracking the bioelectric code as a dynamic control mechanism for determining physical form through pattern-homeostatic processes would include advancement of basic evolutionary developmental biology (Levin and Martyniuk, 2018), as well as numerous biological applications. **My theory is they are missing out the actin cytoskeleton as the central piece of the science puzzle.**

Cracking Nature's Code: In a Nutshell

The actin cytoskeleton provides the centre-piece of the puzzle missing as the Universal architect as the physical form for information of life at its origin, evolution as 'selection of the successful', consciousness, being human and quantum mechanics. Information is conducted from the environment to the cell through the membrane, co-operating with the immune system and mitochondria at the front end, orchestrating a physical process of information triangulation (What If!) and epigenetic control of DNA at the back-end, in a dynamic energetic and physical feedback loop. This acts as a guide to physiological adaptation to the environment (phenotypic plasticity), self-satisfaction and a sense of purpose.

It's Not Just My Imagination!

The actin 'electrome' (De Loof, 2016), which converges science with the spiritual mind-body connection, in conjunction with the plasma membrane acting as the driving force for evolution (De Loof, 2017) supports my theory.

Joining the pieces of my own knowledge and experience, processed through pattern recognition and analogies (diagnostic of the immune system), coincidences (membrane actin detectors) and stretching my imagination (actin networks throw out a wide net of possibilities and fills the gaps) has brought me straight to the point - with actin as the missing piece of the puzzle as the process orchestrator.

Taking an evolutionary perspective on connectivity has been instrumental in allowing a bottom up approach, the systems view from simplicity to complexity. Converging common threads in the process, I propose a unifying process of the origin of life, evolution, consciousness, being human, and the biological interface with quantum theory- the inherent design process- with actin providing the physical form for information.

The universal biology of the process opens its own door to new thinking and opportunities. Current science mysteries, surprises and puzzles reported in the literature may now be better accounted for.

What If! Quantum Effects are Actin's Potential

Quantum effects in various biological systems have been identified with current root cause currently resting on microtubules. Since actin development supports that of microtubules (Colin et al., 2016) it should not be too surprising that actin potentially provides the real answer to the root cause, furnishing simple design principles underlying the control of signalling pathways through self-organization to tune signalling processes.

Electrical Wave Oscillations- Action Potential

Actin's electrical wave-like properties and their action as electrical transmission lines/wires have been reported (Tuszinsky et al., 2004). Of particular interest to them was the travelling wave kink; of particular interest to my mind is the velocity of wave propagation was estimated to range between 1 and 100 m.s–1, which corresponds with action potential velocities in excitable tissues.

Regulation of actin wave generation by calcium (Wu et al., 2012) has allowed me to stretch my imagination, **wonder What If!** to make interesting connections and join some more of the pieces.

Actin structurally informing ATP-ase activation (Murakami et al., 2010) has also allowed me to make a critical connection in its cooperation with mitochondria.

Polarity, Dynamics, Phase Transition, Synchronisation and Information Storage

Actin polymerisation (treadmilling) as a process within a cell is critical to actin's dynamic nature and functionality as tension sensors in its highly diverse applications (Bugyi and Carlier, 2010; Galkin et al., 2012) and holds key pieces of the puzzle: Polarity, speed and design. A paper by Kuhn and Pollard (2005), provides useful data on speed: At the barbed end, the association rate constant for Mg-ATP-actin is 7.4 μM−1 s−1 and the dissociation rate constant is 0.89 s−1. At the pointed end the association and dissociation rate constants are 0.56 μM−1 s−1 and 0.19 s−1. When vitamin D binding protein sequesters all free monomers, ADP-actin dissociates from barbed ends at 1.4 s−1 and from pointed ends at 0.16 s−1.

Self-organised collective oscillatory dynamic behaviour of organisms is universal (Chen et al. 2017b). Actin's role has been identified through self-organised stress patterns driving phase state transitions (Tan et al., 2018) as an explanation for synchronised behaviour from bacteria (Popkin, 2017a) to human crowds (Bain and Bartolo, 2019): It's just a question of scale as a consequence of universal mathematics of life through scaling of connected networks (West, 2014). Fast phase transition is favourable for information storage (Wei et al., 2018).

Quantum Information – Local Action

A comprehensive review which focusses on the physics of the cytoskeleton in terms of physical properties, dynamics, energetics, from an evolutionary (emergent complexity) perspective is fascinating reading (Huber et al., 2013). **I wonder What If!** Actin's unique properties are to conserve quantum energy in universal biological processes, which might explain actin's highly anomalous evolutionarily conserved nature (Galkin et al., 2012).

The key characteristics of quantum cognition bear out the validity of my theory, with membranes, connections and information triangulation (**What If!** ; 'As-If thinking' according to Patalano, 2018) as universal to the process

(Aerts and De Bianchi, 2015) attributed to microtubules (Brooks,2015) rather than actin as root cause.

Quantum Information – Particle Wave Duality-
Spooky Action at a Distance: Magic!

Images of the synapse (Byczkowicz et al., 2018) bear a remarkable physical resemblance to solenoids (https://en.wikipedia.org/wiki/Solenoid). **I wonder What If!** This potentially provides the magic key to opening the door to non-local electromagnetic communication and universal influences and further potential to liven up or defuse the debate on electromagnetic homeostasis organising life (De Ninno and Pregnolato, 2017).This would better account for speeds of reaction to stimuli, brain complexity, underpin quantum mechanics information theory as a gauge as to the measure and value of others and the guide to a progressive evolutionary impulse in conscious evolution.

So exciting is packet- based information communication in the cortex (Luczak et al., 2015): Temporally organised packets of activity lasting ~ 50-200 ms they suggest as basic building blocks of cortical coding. Adding actin's role in driving phase transition (Tan et al., 2018), a unifying theory of brain as a critical system teetering at the tipping point between phases (Oullette, 2018) and treadmilling (Bugyi and Carlier, 2010) to the mix, **I wonder What If! Actin potentially accounts for particle wave duality and the magic key to opening the door of information theory and spooky action at a distance quantum communication.** Facilitated by a conserved hereditable membrane structure (Harold, 2005), common spatial organisation of neural patterns (Chen et al., 2017a) and hereditability of actin networks (Maly and Borisy, 2001; Ktena et al., 2017), further science, pseudo-science and spiritual puzzles start to make physiological sense.

The evidence is really beautifully aligned with quantum information theory (University of Waterloo, 2014) with actin as the potential magic key. **What If! We are the receivers in Universal communication**: Cosmology provides fascinating possibilities, the universe as being fine-tuned for life (Goff, 2018) with actin as a finely tuned dynamic tension sensor (Galkin et al., 2012) as the logical receiver of information to my mind.

Actin Central- Physical Form for Information

Actin is the most abundant protein in the majority of eukaryotic cells (Dominguez and Holmes, 2011) and has been largely overlooked, principally down to its unassuming size and limitations in available scientific techniques. Picking through the science pieces of the puzzle, a common thread is the concept of living organisms as physical mechanical and electrical machines, with actin playing a central role.

Space Time Organisation

A comprehensive scientific review of biological spatial organisation across a wide range of unicellular and multicellular organisms (Harold, 2005) identifies various universal design steps, including molecular self-organization, directional physiology, spatial markers, gradients, fields, and physical forces. The heredity of membranes is described, a key piece of information worth remembering. The cytoskeleton underpins mechanical coupling between distant elements and reaches across the cell membrane to connect with the cell wall or the external environment. It also serves as transport tracks, providing a coherent mechanism of the cell infrastructure linking the membrane and organelles into a fully integrated mechanical system. They leave the big question of where the information comes from to how they all join up in the necessary patterns of organisation open as an urgent life lesson.

Mechanical forces have been acknowledged to play a central role in understanding how biological patterns and morphologies emerge through evolution and how the biological machinery co-operates. A paper by Hernández-Hernández et al. (2014) provides a useful review of background research. Their work provides a comprehensive picture of the cytoskeleton physical forces of tensegrity providing spatial (positional) information as a common framework in both plants and animals. This further reinforces the concept of a physical mechanical machine, with actin playing a central role and a common evolutionary origin.

Membrane Information Receiver – Actin- Protein Plasticity (Conformation) Interactions

The mechanisms behind the cytoskeleton and membrane protein interactions which orchestrate changes in cell shape are reviewed by Doherty and

McMahon (2008). Membrane contact sites with actin also have roles in signalling, metabolism, organelle architecture and trafficking, inheritance, and dynamics (Prinz, 2014; Eisenberg-Bord et al., 2016). A critical feature is coincidence detection. Sub-millisecond coincidence detection has also been identified in dendritic cells (Softky, 1994), further supporting a universal process in a timescale that fits with actin polymerisation. The ATP-ase mediated ability of the membrane to communicate electromagnetic signals energetically at long distances, inside the cell and between organisms has been identified (Tsong, 1989).

Science appears to be switching on to the significance of membrane/protein/ actin interactions with plasticity as a universal energy efficiency mechanism: Mechanotransduction in actin filaments and their role in transmitting mechanical information to protein conformation provides a key piece to the puzzle, for actin central coupling mechanics and complex biochemistry is described by Romet-Lemonne and Jégou (2013; 2016), with exciting possibilities advised for future research. This may have interesting implications for disordered proteins and their plasticity (Katsnelsen, 2017), as a source of on-off switches for all sorts of functions, with interplay between actin and signalling pathways being achieved through simple inherent design principles (Colin et al., 2016).

Changes in structure and motion conspire to shape affinity during the evolution of a protein-protein complex and provide direct evidence for the role of structural, dynamic, and frustrational plasticity in the evolution of interactions between intrinsically disordered proteins (Jemth et al., 2018) indicative of a role for actin in cell phenotype, as the mechanism to direct the construction of cells with the same developmental potential, driven by information received by the membrane. Actin as a central mechanism for communication with the environment to determine and maintain tissue identity and homeostasis in stem cells provides excellent validation (Chen et al., 2018).

Membrane Actin Information Self-Organisation- Pattern Formation- Memory

Actin may solve the mystery of membrane self-organisation and memory (Sezgin et al., 2017), with actin/ protein tethers self-organising biologically important information. **What If! This provides the mechanism for pattern formation for recognition by the immune system** (Pradeu and Vivier, 2016). This is supported by the consolidation of valuable memories during sleep (Antony et al., 2018) whereby information processing occurs in a cyclic fashion during time windows congruous to critical periods of synaptic plasticity (Schönauer et al., 2017) - fits with actin.

Information Coherence

Specificity to stop the wires getting crossed is conferred by catenins, which have emerged as molecular sensors. These integrate cell-cell junctions and cytoskeletal dynamics with signalling pathways that govern morphogenesis, tissue homeostasis, and communication between cells (Perez-Moreno and Fuchs, 2006), ensuring coherence of information.

Actin Mitochondria Cooperation- Is It Worth the Effort?

The diverse interactions between the actin cytoskeleton and mitochondria make a compelling case for a vital connection in the decision process (Jayashankar and Rafelski,2014)- is it worth the effort? Of particular interest to my mind is the actin cytoskeleton is the universal driver of inheritance of the fittest mitochondria a vital decision process as to divide or die (Boldogh and Pon, 2006). Actin releases energy from mitochondrial ATP using ATP-ase (Kuhn and Pollard, 2005). Actin structurally informing ATP-ase activation (Murakami et al., 2010) has also allowed me to make a critical connection in its cooperation with mitochondria.

Mitochondria and Immune System Cooperation – Amplification- Adaptive Cascades

Mitochondria act as a platform for the innate human immune response (Foley, 2016) which operates through a change-detection system whereby it recognises the presence or absence of patterns as a unifying framework (Pradeu and Vivier, 2016). The immune complement system acts as a natural amplification system (Janeway et al., 2001). As a consequence, the activation of a small number of complement proteins at the start of the pathway is hugely amplified by each successive enzymatic reaction, resulting in the rapid generation of a disproportionately large complement response and adaptive cascades creating an information amplification system.

Information Weight Evaluation- Synaptic Plasticity

Actin has the potential to provide a simple solution to synaptic plasticity (Ponte Costa et al., 2017): Actin engineers the synaptic gap (Byczkowicz et al., 2018) with synaptic weighting as a consequence of its physical tensegrity and electrical properties determining the energy value of information, not necessarily just the amount of neurotransmitter (Wang et al., 2018). Actin has been identified as the universal information processor for detecting dead cells, identifying them for disposal by the immune system (Ahrens et al., 2012). The immune system targets dead cells through selection/sensing of exposed actin (Brown, 2012), with pruning of unused synapses during sleep representing the dynamic feedback loop (Paolicelli et al., 2011) to retain valuable and dispose of redundant information.

Information Triangulation- What If!

Triangulation (**What If!**) plays a universal role in natural and artificial networks (Gorochowski et al., 2018), transforming 2D information to 3D: Grid cells firing together through triangulation provides the ability to manage spatial organisation for internal memory networks (Doeller et al., 2010) and more complex 7D information geometry across the brain (Ananthaswamy, 2017). The key characteristics of quantum cognition bear out the validity of my theory, with membranes, connections and information triangulation as universal to the process (Aerts and De Bianchi, 2015). A feedback loop from glial cells to orchestrate synaptic plasticity has been proposed (De Pittà et al., 2016).

DNA Dynamic Activation- Methylation Patterns

A review of actin's role in transferring mechanical information, which is translated into biochemical signals across the entire cell from plasma and nuclear membranes in stem cell regulation makes very interesting reading (Uzer et al., 2016). The existence of an actin nucleoskeleton has been hotly debated and relatively recently emerged (Galkin et al., 2012). **I wonder What If!** The physical twisting of the DNA blueprint, orchestrating gene expression (Irobalieva et al., 2015) is attributed to actin. Its direct impact on gene expression in stem cell transformation into bone has been identified (Sen et al., 2015; Chen et al., 2015).

The dynamic active nature of DNA methylation patterns has only recently been identified. A review (Schuermann et al., 2016) considers its role in embryology, cell differentiation and gene expression, activities physically managed by the membrane orchestrated by the actin cytoskeleton (Perez-Moreno and Fuchs, 2006). Bacterial (Pacis et al., 2015) and viral infection (Lichinchi et al., 2016) directly impact methylation patterns, most likely orchestrated by actin since it is universally hi-jacked in viral infectivity (Cudmore et al., 1997; Ohkawa and Volkman,1999 ; Lu et al., 2004; Marek et al., 2011).

DNA Valuable Information- Actin Turn On

Valuable information is secreted in 'Junk' DNA: Junk DNA is not really junk after all (Kershner, 2013). It makes up 95% of DNA and contains high security valuable information involved in regulatory evolutionarily significant processes (as opposed to protein coding). This article provides a good introduction as to how junk DNA is involved in the orchestration process (Barras, 2016). The ENCODE Consortium in the UK has identified all-sorts of previously hidden switches, signals and signs throughout the entire length of human DNA (Encode, 2012), likely switched on by actin.

DNA Long Term Mutation

DNA based evolutionary change is likely captured when dynamic continuous adaptation has proved successful, with gene splicing for efficiency (Iñiguez and Hernández, 2017) likely under instruction from actin.

Dynamic Feedback loop

The dynamics and direction of the adaptation process appears to be proportionate to environmental pressure. Symbiotic ant-plant partnerships have been established as having rapidly evolving genomes to reflect the active co-operative synchronisation with changes in environmental pressure (Pennici, 2016). Successful predator- prey relationships and ecosystem dynamics in space and time require prediction of environmental temperature cues (Neuheimer et al., 2018).

Reverting to previous phenotypes under environmental pressure reinstates previously more successful design adaptations, is also reflective of the reverse feedback loop as demonstrated in carp (Panko, 2016). Potential skin surface design selection appears to be a matter of pattern selection (Cooper et al., 2018) as a consequence of memory through membrane/ actin self-organisation.

Original Architect of Life's Design

The story of the origin of life is a classic one of chickens and eggs, with a lot of hot debate. That protein came first is winning (Cepelewicz, 2017a), not ribonucleic acid(RNA). RNA lacks the required synchronisation to maintain the symphony of chemical reactions (Cepelewicz, 2017b) with synchronisation and information processing being a universal function of actin.

Universal design principles of tensional integrity or 'tensegrity' coined by Buckminster Fuller (http://www.scholarpedia.org/article/Tensegrity), including nature's geodesic design geometry of architectural form at the origin of life are described by Ingber (2000): The focus of evolutionary biology on genetic information processed by natural selection, is challenged with a simple step wise scenario (Occam's Razor). **All the pieces of the puzzle can only fall into place when the cytoskeleton and metabolism are both taken into account**. Information processing through tensegrity of the cytoskeleton at the origin of life is further elaborated by Ingber (2003).

Phosphite is believed to have played an important part in primordial life (Tapia-Torres and Olmedo-Álvarez, 2018), with actin structurally informing adenosine triphosphate (ATP)-ase activity (Murakami et al., 2010) providing further evidence for natural selection of actin as the architect of choice.

The cytoskeleton was considered to be the natural preserve of multicellular organisms until quite recently (De Souza, 2012), previously creating an evolutionary disconnect. Comparison of actin and myosin in bacteria, archaea and eukaryotes supports actin as the universal architect (Aylett et al., 2011).

Strong evidence for an identical metabolic framework in all living organisms, compliant with the design principles of robust and error-tolerant scale-free networks, may represent a common blueprint for the large-scale organization of interactions among all cellular constituents, is provided by Jeong et al. (2000). That the actin cytoskeleton is the missing piece of the puzzle is further supported by Jockusch and Graumann (2011) and Wickstead and Gull (2011).

Evolutionary Driver

Evidence provided here potentially defuses the surrounding hot debate as to how cells engineered the transition to multicellularity (Singer, 2015). Symbiosis is proposed as a general principle for evolutionary innovation (Aanen and Egleton, 2017). Wickstead and Gull (2011) identifies an original primitive actin cytoskeleton in the design, responsible for importing symbiotic bacteria in the key transition to eukaryotic cell evolution, through synchronisation (diagnostic of actin). Additional evidence for a 'master developmental program' and failing to divide as a simple pathway to multicellularity (Dayel et al., 2011) also accounts for the remarkable similarity in embryo development across species, with actin as the driving force (Frum and Ralston, 2018). Actin's instrumental role in control of cell division (Field and Lénárt, 2011; Mogessie and Shuh, 2014; Kaur et al., 2014; Baarlink et al., 2017) and orchestrating cell differentiation has also been demonstrated in stem cell research (Sen et al., 2015).

A singular symbiotic event with membrane bioenergetics has been proposed to holding the key to complexity (Lane, 2017) ; **I wonder What If!** Actin might hold the key. The CoRR (co-location for redox regulation) theory of bioenergetics proposes an explanation as to why genetic information is retained in organelles (e.g mitochondria and chloroplasts, as a consequence of symbiotic adoption of prokaryotes) in eukaryotes has a selective advantage of subcellular co-location of specific genes with their gene products (Allen, 2017).

I wonder What If! It could be the retention of an intimate energy efficient framework afforded by actin that has been vital to evolutionary success to synchronise the metabolic process. This explains the absence of transfer of genetic information to the host genome expected from symbionts (Keeling and McCutcheon, 2017) which totally misses the point of actin.

Human Brain Evolution

Universal events have been shown to be responsible for evolutionary leaps in human evolution of intelligence, according to the Pulse Climate Variability theory which provides a strong conceptual framework within which to examine other evolutionary events (Maslin et al., 2015). The structural complexity of the brain has a universal part to play, with the extraordinary uniformity of the 6 layered structure of the pre-frontal cortex responsible for the superior human executive decision making process and more complex 7D information geometry (Ananthaswamy, 2017) allowing superior human perception and imagination. With superior pattern processing as the essence of the evolved human brain (Mattson,2014) and actin as the evolutionary driver, it should have come as no surprise that the immune complement cascade (Janeway et al., 2001) is emerging as the architect of the developing brain (Coulthard et al., 2018).

Quantum Leaps - Universal Events Driving Evolutionary Epochs

With the same universal process at play, I support the question **What If!** Life was easy at its origin (Sarchet, 2016), casting doubt on the theory that hot volcanic vents were required to break through the energy barrier for the transition to eukaryotes (Sarchet, 2018).

Actin's explosive nature as the universal force provider in eukaryotes (Xue and Robinson, 2013) provides fascinating possibilities to my mind as to the root cause for other evolutionary quantum leaps (Epochs) to deliver Epochalyptic events such as the Cambrian explosion caused by universal events.

Actins Anomalies Naturally Selected

With new understanding of a universal architecture, brings new insights and opportunities. A comprehensive review which focusses on the physics of the cytoskeleton in terms of physical properties, dynamics, energetics, from an evolutionary (emergent complexity) perspective is fascinating reading (Huber et al., 2013).Particularly attractive to me, it throws another spanner in the DNA wheel, focuses on actin's superior physical properties for adaptiveness (apparently expensive energy- wise but evidently worth it other-wise!) its pervasive distribution and central role in morphogenesis lending itself as a key driver of evolution and consciousness. The paper also refers to quorum sensing as a mechanism to drive cooperation and efficiency as an evolutionary advantage, another key piece of the puzzle, but misses the point of actin.

I wonder What If! Actin's unique properties are to conserve quantum energy in universal biological processes, which might explain actin's highly anomalous evolutionarily conserved nature (Galkin et al., 2012). **I wonder What If!** Actin as the answer to the puzzle can be derived from the machine learning approach to physical properties of materials and molecules (Bartók et al., 2017). The universality and the systematic nature of their framework provide new insight into the potential energy surface of materials and molecules.

Natural Selection of the Successful

Actin as the universal physical form for information provides the foundation for universal communication, co-operative physiological dynamic adaptive evolution, according to the electromagnetic laws of attraction and resistance that is natural selection. Dynamic epigenetic control of nuclear DNA is orchestrated by actin and provides sufficient speed and direction of adaptation, proportionate to the environmental pressure for evolutionary success.

Information transfer, evaluation and decision making, is afforded by communication within and between cells, universal to natural selection. The process is more reflective of 'selection of the successful' rather than 'survival of the fittest' which should provide a healthier perspective on the current human constructs and makes much more sense to my mind than random genetic mutation.

Essence of Life- Effort and Value- Is it worth IT?

The essence (energy sense) of life, to select life or death is the ultimate decision (Inzlicht et al., 2018), influenced by the value and return on investment and associated dopamine reward (Walton and Bouret, 2018). The diverse interactions between the actin cytoskeleton and mitochondria in multicelled organisms make a compelling case for a vital connection in the decision process (Jayashankar and Rafelski,2014). Of particular interest to my mind is the actin cytoskeleton is the universal driver of inheritance of the fittest mitochondria (an alternative much better fit with actin as the driver of natural selection of the successful rather than survival of the fittest), a vital decision process as to divide or die, engineered by actin (Boldogh and Pon, 2006): Is life worth IT?

Bacteria Naturally Selected – Risk Decision Process and Immunity of Mitochondria

Quorum sensing has been recognised as the universal risk decision making process in bacteria as a mechanism for kinship (Schluter et al., 2016), as an orchestrated response (diagnostic of actin, only recently identified in bacterial cells, previously an evolutionary disconnect in the actin central story; De Souza, 2012). The evolution of quorum sensing for bacterial communication as an early step in the development of multicellularity has been proposed (Miller and Bassler, 2001). A review by (Popat et al., 2015) provides a useful background, focussing on cultural implications of establishing a natural homeostatic balance- effectively acting as an 'immune' self-protection mechanism, with co-operation and symbiosis as the ultimate objective. Interesting analogies with what is wrong with human behaviour in our current cultural mix.

Social networking is a natural phenomenon in bacteria (Remis et al., 2014). The extracellular matrix of bacterial biofilms is much more than passive glue. It provides a medium for long range electrical communication (Prindle et al., 2015), as well as other dynamic functions between cells (Dragoš and Kovács, 2017) to organise and synchronise activities. An article by Popkin (2017b), beautifully illustrates the synchronisation in time-lapse videos.

The ability for individual 'simple' microorganisms to collectively co-operate and self-organise as part of the evolutionary transition to multi-cellular organisms is reviewed in a paper by Claessen et al. (2014). They discuss the success strategies that are used by bacteria (biofilms, filaments and fruiting bodies) to achieve what can amount to extraordinarily sophisticated physical structures, particularly under stress, clearly demonstrating inherent design. They provide insights on the universal involvement of pattern recognition (diagnostic of the immune system ; Pradeu and Vivier, 2016) and orchestration with apparent wave like co-ordinated action (diagnostic of actin Wu et al., 2012) that provide key clues to the connectivity of the underpinning process. A universal evolutionary process is recognised as a future exciting development – Not something that they offer. How exciting!

Failure to select a successful co-operative environment results in extreme behaviour even in bacteria. When threatened by a harmful organism invading their space they can revert to powerful self-assembling energy expensive lethal weapons resembling an elaborated actin barbed end (Le Page, 2016). More analogies can be usefully learned in human behaviour.

Life Form - Design Integrity – Immune Pattern Recognition- Mitochondrial Cooperation

Actin's universal role in selecting the nature of a cells life form has been demonstrated in stem cell differentiation (Sen et al., 2015: Chen et al., 2018) and embryology (Frum and Ralston, 2018)- enough to illustrate. Support for the idea that actin cooperates with immune design integrity operating on a mitochondrial platform - our internal police force keeping order from potential chaos is provided by Bement and von Dassow (2014): Pattern formation and dynamics of actin cytoskeleton were identified as universal to cell design, the order and significance of actin being much greater than generally recognised.

Friend or Foe – Selection or Resistance

Environmental information received by the membrane is translated by actin to effect diverse physical reactions including immune synapse development (Chaabra and Higgs, 2007), organise essential processes such as endocytosis (Bezanilla et al., 2015), key processes involved in natural selection of symbionts and resistance to bacterial pathogens (Weiner and Eninger, 2018) to determine susceptibility to disease and its consequences.

There is a remarkable similarity between the bacterial barbed spear and the spear used by bacteriophages (bacterial viruses) to penetrate bacteria instrumental in the CRISPR technology, the origin of which is an evolutionary mystery (Ledford, 2017): **Or is it just my imagination?**

Insertion of the viral genome confers immunity from further attack. The membrane- actin- DNA connectivity and immune system cooperation could provide an explanation. This could also explain the cross-species jumping as a viral evolutionary success strategy (Geoghegan et al., 2017) to avoid detection and allow invasion.

Viruses communicate between themselves in order to decide as to their course of action (Erez et al., 2017) and operate in collective infectious units to improve their chances of success (Sanjuán, 2017) with actin communication as likely root cause. The actin cytoskeleton has at least been shown to have a universal role in viral infectivity (Cudmore et al., 1997; Ohkawa and Volkman,1999 ; Lu et al., 2004; Marek et al., 2011). The explosive nature of actin (Xue and Robinson, 2013) has been capitalised to provide a projectile mechanism to exit the host (Newsome and Marzook, 2015).

Sense of Direction and Purpose- Networking

Networks are universal in biological and social systems. Success requires a complex efficiency evaluation as beautifully demonstrated by slime molds (Tero et al., 2010).

Attraction and resistance as the guiding principles in selecting directional development in nerve axon networks (de Ramon Francàs et al., 2017), with actin controlling axon growth cone formation (Blanquie and Bradke, 2018) lending strong support to universal electromagnetic homeostasis controlling life (De Ninno and Pregnolato, 2017) with actin as the universal architect.

Consciousness- Sense of Purpose- Electrome Vibration

I wonder What If! Consciousness is universal by inherent design. Science is waking up to the reality (Chittka and Wilson, 2018), that it's not just a human construct (Holmes, 2017). The 'Orch OR' theory of human consciousness is reviewed together with and responses to their many critics make interesting reading (Hameroff and Penrose, 2014). Stuart Hameroff is an anaesthesiologist. He might be surprised that plants are equally affected.

Plants having their own neurobiology has been on the radar for a while and consciousness tentatively suggested (Ananthaswamy, 2014). Science has started to attempt to tap into the language of plants based on electro-chemical signalling (Uys, 2016) opening the door to novel sustainability solutions. Consciousness of plants has been better substantiated by demonstrating a risk response (Dener et al., 2016; Morell, 2016), the process of which they are not aware. A 'remarkable' orientation of roots towards a cathode or anode (Kral et al., 2016) is no surprise to me. As far as I knew at the time of stretching my imagination in developing my theory, no one else had made the same universal connections.

I have since discovered the 'electrome' which validates my theory (De Loof, 2016). The demonstration of the impact of anaesthetics on plants blocking action potential (Yokawa et al., 2017) is rather beautiful to me; the authors are unclear as to the mode of action. Vibration as a successful strategy for communication and selection has been identified between plants and animals (Clarke et al., 2017) and between animals (Hill, 2015).

What's the Point of Consciousness- Sense of Self, Purpose and Happiness- Good Vibrations!

I wonder What If! the point being a sense of self, as a guide to achieving self-satisfaction and finding a sense of purpose and direction in life through valuable work, valuable relationships and happiness - Good Vibrations!

The same idea of purpose as the point is supported by Sir Roger Penrose, accepting attribution to microtubules are very likely to be incorrect as the root cause (Paulson, 2017).

The likely overlooked role of actin is supported in the pieces I have connected in its potential role in the origin of life and as supported by Tuszinsky (2014), who predicted actin bridges the gap with the 'Orch OR' theory of quantum consciousness and the potential for monumental breakthroughs. Hopefully my monumental effort will be worthwhile.

Sense of purpose or direction is recognised as key to health and happiness, but current theories as to the underlying process lack substance (Burrell, 2017). This article suggests if it turns out to be a physical issue you would see lots of funding going into it- I sincerely hope so.

No Wonder Nature Selected Actin as the Universal Architect!

Being Human by Design

We Are Only Human- Universal Information Processing

The following pieces of the science puzzle to my mind make a strong enough case to illustrate that we are only human as a simple more energy efficient elaboration (energy- work-action) of the universal information processing process: Actin's specific role in individual neurons (Dance, 2016); space time organisation mechanism of the connectome during brain development (Kaiser, 2017) ; membrane, actin, mitochondrial co-operation in neuronal spatial organisation (The Royal Society, 2017); sub-millisecond coincidence detection and associated depolarisation in dendrites (Softky, 1994) ; mitochondrial cost-benefit/risk decision process (Popat et al., 2015) ; actin's role in driving axon growth cone (Omotade et al., 2017); actin's electrical wave propagation speed corresponds with action potential (Tuszinsky et al., 2004) ; neural synapse orientation management (Mayo and Smith, 2016) and actin organisation of exocytosis and endocytosis across the synapse for information transfer (Maritzen and Haucke, 2018).

Brain Design- Superior Pattern Processor

Superior pattern processing is the essence of the evolved human brain (Mattson, 2014) as a consequence of the immune actin interaction I propose at its evolution. This is afforded by an extraordinary uniformity of the 6 layered structure of the pre-frontal cortex responsible for the superior human executive decision making process allowing superior human perception, imagination and abstract thinking (Dumontheil, 2014).

Vision - Case Study- Actin Fills the Gaps

Vision is just another information processing system (Diamant, 2008) with the same universal physical properties which recognises shape and spatial organisation from past experience. A seamless visual experience from actin filling the gaps in the blind spot based on expectation appears reasonable to me (He and Davis, 2001).

Beautiful to me was finding a paper on the development of the structural complexity of the eye. Development of vision photoreceptor cells has been identified as being orchestrated without genetic influence by self-organisation of the cytoskeleton. It took 30 years of research to exhaustively search for a genetic blueprint without success (Kumaramanickavel et al., 2015).

Information processing is separated into layers within the visual cortex in response to environmental stimuli (Muckli and Petro, 2017) and adaptation occurs through space –time recurrent connectivity giving a network a powerful and potentially multipurpose ability to compute complex functions of its information inputs.

Tuning takes place in timescales of at least several hundreds of milliseconds, in keeping with new information input from eye movement (del Mar Quiroga et al., 2016), a timeframe reflective of actin mechanics. A unifying mathematical framework of cortical activity, linking network structure and function demonstrates emergent electrical activity mirrors its hierarchical structural organization (Reimann et al., 2017), reflective of actin.

Actin synchronisation of vision with hearing (Gruters et al., 2018) provides a foundation for information triangulation to make the best sense from the information input as to the reason why (Woodward, 2017). The colour piece of the puzzle is still not clear. Some additional light is cast on the matter when considering the colour blue: Language has a direct impact on colour perception (Winawer et al., 2007). If there is no word for it then you have no perception of it. **I wonder What If!** In the absence of a pattern or coincidence to be synchronised with you don't recognise it.

Brain Network Complexity –Intelligence

Glial cells are extraordinarily important players, forming 90% of the brain-not just neurons by any stretch of the imagination. A review by Araque and Navarette (2010) give a fascinating overview: These include key clues as to how they communicate in a calcium dependent wave-like fashion (diagnostic of actin) that co-operate as a neuron/glial network, universal to the process. Synchronisation of microglial cell development is critical in order to perform different functions to accommodate temporally changing needs by active engagement in synapse pruning and neurogenesis, then maintaining homeostasis associated with the immune response (Matcovitch-Natan et al., 2016).

I wonder What If! This is most likely to be orchestrated by actin, as for other synchronised activities. This is supported by microglia being responsible for synaptic pruning in brain development (Paolicelli et al., 2011), sensing actin as the root cause (Brown, 2012).

Energy efficiency through elaboration of the universal process into the complexity of the brain is achieved by synchronisation to encode and passage information via rhythmic electrical activity patterns between regions (Tingley et al., 2018). Triangulation (What If!) plays a universal role in natural and artificial networks (Gorochowski et al., 2018), transforming 2D information to 3D: Grid cells firing together through triangulation provides the ability to manage spatial organisation for internal memory networks (Doeller et al., 2010) and more complex 7D information geometry across the brain (Ananthaswamy, 2017) with modularisation with functional specialisations as geometric polyhedra (Genon et al., 2018).

The Network Neuroscience Theory of Human Intelligence (Barbey, 2017) describes a variety of attributes that support actin as the universal architect of the design and efficiency : General intelligence, g, emerges from individual differences in the network architecture of the human brain (membrane/ actin inherited self), crystallized intelligence engages easy-to-reach network states that access prior knowledge and experience (memory based) and fluid intelligence engages difficult-to-reach network states that support cognitive flexibility and adaptive problem-solving (through pattern recognition of analogies) and **wonder What If!** Actin fills in the gaps to stretch the imagination, as is likely in filling the vision blind spot.

Actin' Badly

The impact of mild stress conveying biological survival benefits has been recognised for some time. Research on the roundworm *Caenorhabditis elegans* (Kumsta et al., 2017), selected for study for its natural transparency, illustrates the positive impact stress has on breaking down protein aggregates. They identify exciting potential benefits to understanding the biological basis might find solutions to neurological diseases such as Huntington's, Alzheimer's and Parkinson's, which are similarly caused by aggregating proteins. The authors wonder what the cellular memory is (Sanford-Burnham Prebys Medical

Discovery Institute, 2017). They took a complex genetic approach. Judging by the images in the paper, **I wonder What If!** Is it as simple as causing the actin cytoskeleton to stretch out, unblocking mechanical and energy barriers to facilitate wave and information flow, removing blocks to success and encouraging action – or is it just stretching my imagination too far? I have since found evidence - Kurakin and Bredesin (2015) were already on the case, with actin as root cause of Alzheimers and other degenerative disorders and diseases such as cancer.

Life could be a lot easier if we used our AI technology tools to best effect. Joining the pieces of the science information puzzle and validating What If! imaginative hunches, using AI revealed the central role of actin in degenerative disorders such as Alzheimers and cancer (Kurakin and Bredesen, 2015) and would avoid picking through the pieces less painfully than I have attempted to do in a more exhaustive and less exhausting manner. Then what is the point for the research to be buried back in the literature?

Universal Functional Process

Evidence for the universal cellular process elaborated as the root cause of a selection of functional processes is provided:

Working memory is the functional framework underpinning high-level cognitive processes where patterns of dynamic connectivity have been identified as being universal (Stokes, 2015). Its distributed non-localised nature reflects a generic mechanism to transform information input (Christophel et al., 2017). Shared mechanisms for spoken recall (Patai and Spiers, 2017) and the way in which individuals remember the same event (Chen et al., 2017a) reveals common spatial organisation to support a common framework of neural patterns. This also supports the critical nature of a conserved hereditable membrane structure (Harold, 2005), evolutionary conservation of actin (Galkin et al., 2012) and hereditability of actin networks (Maly and Borisy, 2001) in the evolutionary process and implicit in the evolutionary membrane memory exhibited during embryology and hidden treasures buried in 'Junk' DNA.

Emotion: Memory control has been identified as a universal control of emotions (Engen and Anderson, 2018). Emotion is a universal process not localised (Wager et al., 2015). Patterns of electrical brain activity have been identified in cracking the emotional code (Coghlan, 2018). Connectome harmonics of brain activity are spatial patterns of synchronous activity emerging on the cortex for different frequency oscillations and functional connectivity patterns, responsible for emotional reactions (Atasoy et al., 2017) and action stop signs (Duque et al., 2017), triggered by the surprise of predicted negative consequences (Wessel, 2018). Facial expression provides an emotional feedback loop as a tool for social influence (Crivelli and Fredlund, 2018), with facial plastic surgery having unsurprisingly negative consequences on children's behaviour (Personal observation!).

Emotion Contagiousness and Empathy: The word contagiousness is normally associated with spread of disease as a consequence of the immune system. Transmission of emotional feelings can equally be effected through non-physical routes, both negative (Cold: Cooper et al., 2014; Pain : Smith et al., 2016) and positive (Berdahl et al., 2017) through synchronisation for adaptive social behaviour. Emotion contagion is thought to have an evolutionary foundation as the basis on which empathy is built (de Waal, 2008).

Learning: Another universal process based on experience in organisms of all complexity. Science is starting to dis-entangle how our brain manages the process. Key clues for a universal role for actin – membrane interaction is based on an article by Fields (2016) : The first wave of pre-conscious perception (absorbing information subconsciously) takes 160ms (perfect timing for actin to get its act together) to establish patterns and coincidences of previous knowledge/information before the second wave where new information is incorporated.

Language learning: Noam Chomsky's computational theory of universal grammar in the 1960s is built on the foundation of a brain wired with a mental template for grammar. Birds learning to sing provides interesting clues to language learning through convergence of patterns of sequences (James and Sakata,2017) and supports recent theories where general cognitive abilities of word association analogies (an-allergy) and predictive capabilities are at play (Ibbotson and Tomasello, 2016). Hence things get funny as a consequence of a train of thought travelling in an unexpected direction! A theory of experience input, patterns and imagination output are entirely complementary to my theory (Dor, 2015).

Sensational Supramodality: All forms of sensory information processing are ultimately based on decoding the space and/or time structure of incoming patterns of action potentials (Karmarker and Buonomano, 2007). A universal process has been demonstrated to operate for all senses- supramodality- as revealed by electrophysiological recordings (Faivre et al., 2017). Human experience is an imaginative perception of sensory information (de Lange et al., 2018) with actin most likely making best sense of the senses. It's sensational - when attention is otherwise focussed and you are not expecting it – what do you see? (https://www.youtube.com/watch?v=vJG698U2Mvo). Lack of synchrony explains the rubber arm illusion (https://www.youtube.com/watch?v=sxwn1w7MJvk; Bekrater-Bodmann et al., 2014).

Decision Making: The mind as a decision-making organ as the organizing principle of psychology has been identified (Gintis, 2007). Mental maps also have a key role (Kaplan et al., 2017), with value guiding behaviour (Davidow et al., 2018) and personality (sense of self) as a space gradient in the prefrontal cortex (Sul et al., 2015). **I wonder What If! The universal process I propose provides the mechanism they have missed and has an excellent fit with elements of the unified theory of decision making** (Ashtiani and Azgomi, 2015) including environmental triggers, cultural cooperation, emotion, action potential, risk assessment and cost-benefit. The decision process in the event of parallel conflicting memories is determined through weighting of synapses (Felsenberg et al., 2018), as measured by actin.

Stretching the Imagination: The strong correlation between predictive spatial cognitive ability with creativity and innovation is just starting to be recognised as a significant untapped source of human potential (Clynes, 2017). This has a perfect fit with the mathematical model of the evolutionary process for emergent novelty, innovation and breakthrough (Loreto et al., 2016), where the brain makes connections with existing knowledge to create ideas by exploring the adjacent possible (diagnostic of the connectome gaps). To my mind this essentially describes the imagination process. Using my active imagination, **I wonder What If! Joining the pieces in the inherent design process creates an image/ idea. Decision making rejects or accepts**

the idea, based on risk / cost-benefit. Acceptance is followed by validation through repeating the process before taking action.

The more far-fetched the idea (beyond self- knowledge and/or accepted wisdom) the more validation and evidence is required, to ensure feeling safe not foolish before asking the 'stupid' question, challenging the *status quo,* breaking with cultural/accepted norms and taboo and causing an immune reaction from your audience - literally **stretching the image-in-actin.**

Sleep and Dreams : The connection between sleep and vitamin D has been identified (de Oliviera et al., 2017), and vitamin D involved in the actin polymerisation process (Kuhn and Pollard, 2005) with actin's role in synaptic homeostasis associated with slow wave oscillations apparent in sleep (Tononi and Cirelli, 2003) highly likely. Pruning of unused synapses during sleep (Paolicelli et al., 2011) to retain valuable information is achieved by the immune systems attraction to actin (Brown, 2012). Severe mental health issues are apparent as a consequence of process failure, such as Alzheimer's, for which actin has been identified as root cause (Kurakin and Bredesin, 2015). The impact of acute sleep loss on DNA methylation and metabolism (Cedernaes et al., 2018) also supports actin as the root cause. **I wonder What If! Dreams are actin filling in the gaps between imagination and reality, as for vision in the real woken world! The 'Qualia'of Quantum effects.**

What Not Genetic! Kind by Nature

Genetics is struggling to explain the uniqueness of being human (O'Bleness et al., 2012), let alone the uniqueness of being yourself. Networks of information provide the universal building blocks of human experience (Falk and Bassett, 2017). The confluence of understanding of networks in neuroscience and social networks are beginning to shed light on how ideas and behaviours are spread. It shouldn't really be a surprise that we are Human Kind by nature (Rand et al., 2012), naturally selecting real friends by tuning in (Parkinson et al., 2018) and synchronising (Lieberman, 2018) with those on the same wavelength (Anders et al., 2016) through quantum cognition. Evolution of cooperation in Game Theory through thermodynamics and magnetic alignment of players (Adami and Hintze, 2018) and reciprocity (not money!)

at a global scale to stabilise cooperation (Frank et al., 2018) should make the world go round – once the universal biological process is better understood.

Consciousness- Deciding What's the Point?

The fundamental process is deciding whether it's worth the effort of taking action to serve a valuable purpose.

I wonder What If! The 'mystery' 200ms gap in human consciousness between making a decision and taking physical action (Williams, 2013) can be accounted for by actin and mitochondrial co-operation: Actin's electrical wave-like properties and their action as electrical transmission lines/wires whereby the velocity of wave propagation was estimated to range between 1 and 100 m.s–1, which corresponds with action potential velocities in excitable tissues (Tuszinsky et al., 2004). Actin's polymerisation rates at the barbed end, the association rate constant for Mg-ATP-actin is 7.4μM–1s–1 and the dissociation rate constant is 0.89s–1. At the pointed end the association and dissociation rate constants are 0.56μM–1s–1 and 0.19s–1 with actin releasing energy from mitochondrial ATP using ATP-ase (Kuhn and Pollard, 2005). This fits the timeframe for actin to get its act together.

Without hope, a forward direction and sense of purpose, 'Give-up-itis'sets in (Leach, 2018) : Deciding what's the point?

Inherent Design- Actin's the Point

An analogy between learning theory and evolution has usefully hit on an inherent design mechanism (Watson and Szathmáry, 2016) with focus on genetic networks and the point of actin is totally missed. Neuroscience is puzzling over whether it is limited by tools or ideas. Fragments of a "theory of design" are described, to uncover the underlying simplicities (Mitra, 2017). Equally, science knows shockingly little about the origins of electroencephalography (EEG) signals (Cohen, 2017). My ideas of inherent design with actin the universal architect appear to point in the right direction.

In Essence- Unique Sense of Self- IT Factor

In essence your membrane actin interaction IT factor provides your unique sense of self and provides inner guidance to success, self-satisfaction and a sense of purpose- Happy Days!

Pineal Gland- A Special Case- Science Meet Spirituality

Like your eyes, the pineal gland has the same architecture of rods and cones to process light despite being smack bang in the middle of our brain. According to ancient wisdom, the pineal gland is recognised as the seat of consciousness (Harper, 2016), otherwise known as the mind's third eye (real-eyes) Chakra- a mostly untapped human resource. This connects with studies which demonstrate differences in personality traits between more or less open minded creative people, with more positive behavioural traits apparent in the former.

Having an open mind and paying attention are associated with creative thinking (Kreitz et al., 2015) and merging separate images presented to each eye (Antinori et al., 2017) connect in my mind's eye to the ability of the eyes to triangulate with the pineal gland. The connection with consc(ient)iousness is particularly beautiful.

Opening the mind's eye is otherwise achievable through meditation or taking of psilocybin— a hallucinogenic compound naturally found in magic mushrooms, structurally similar to serotonin, altering the filter of consciousness (Klein, 2017) - the scientists would like to know how? **I wonder What If! Triangulation with the pineal gland is likely to be the root cause.** The neural patterns visualised under the influence of psychedelic drugs (Coghlan, 2017) appear to support the idea. Validation comes from bird brains (Cassone and Westneat, 2012) where the pineal gland is instrumental in guiding direction of migration through triangulation.

Science Meet More Spirituality- Being Mindful

Progressively, spiritual phenomena long recognised in Ancient Wisdom and Eastern medicine such as acupuncture (Walia, 2016), hypnosis (Dahl, 2016), yoga (Pelechowicz, 2016) and energy healing (Hammerschlag et al., 2014) have been at least vindicated if not fully embraced by science. The interstitium (Olen, 2018) as a 'new' organ has recently been labelled by science- it appears to be what has been described as Qi meridians for quite some time. The crown chakra also has a new scientific identity (Miller et al., 2018). Some if not all can be directly attributed to the central role of the actin cystoskeleton.

Super Sensitivity Side Effect

From personal experience, an unexpected consequence of actively paying attention to the inherent design process, consciously on the lookout for information pieces to connect, coincidences and analogies acts as a creative switch to an open (growth) mind-set. My sensitivities to other people's feelings and motives, acuteness of hearing, intuition and sense of purpose have all grown enormously- with some euphoric natural highs thrown along the way to keep me on track. It might not be sensible for everybody to wake up at the same time.

We Are Only Human After All!

Game Changing - Bridging the Gaps - Realising Full Potential

Cracking Nature's Code- Bridging the Science Gap to Change the Game

I share the sentiments of Gurel et al. (2014) who have joined the pieces connecting the actin cytoskeleton to the endoplasmic reticulum (ER) and golgi body: 'A tendency in cell biology is to divide and conquer. We hope that this review helps to clarify our current understanding ..., by inserting new pieces into the puzzle. While some of our pieces might be inserted somewhat incorrectly, we hope that their presence will allow others to replace them with better-fitting ones.'

The gap I have managed to bridge couldn't have had a much wider span if I had tried. I have started to assimilate evidence on physics, Universal influences and space time, but resisted the temptation 'to boldly go where no man has gone before' not feeling enterprising enough (LOVE Star Trek – Science faction is the way to go). Fortunately I have found somebody else that has already started to go there. My endorsement from William Brown speaks volumes (see below). Also, the Mind to Matter book (Church, 2018) gives chapter and (Uni)verse on the science and symptoms of our amazing creative potential and what a game- changer it could be if it was harnessed. The science of the root cause rests on microtubules with focus on the potential: 'There is no proof supporting the theory that consciousness lives inside the brain... Despite the lack of evidence at present, however, materialist sceptics assure us that science will eventually fill in the gaps '. My piece of the puzzle surely does that and this book hopefully paves the way for its scientific acceptability and my technology idea has the potential to facilitate and accelerate the process of Changing the Game. Perfect timing!

Science Puzzles Solved? Potentially!

Tom Siegfried (2015), acclaimed science author and proponent of the radical new physics of information has set out his top 10 science puzzles for the 21st Century. Actin as the universal architect of life as information that I propose, goes a long way towards answering No.10: How did life originate- The universal biological physics/mechanics process underpinning life's origin and explaining the strategy of game theory; 6: Genes, cancer and luck- Cancer caused by failure of control of the actin information process; 4: The nature of time - Perceived time as a product of the temporal function of the membrane and actin cytoskeleton connections and 1: The meaning of quantum entanglement- Quantum entanglement connected with actins phase transition and other physical properties. **(Not all evidence is presented here due to running out of space and time).**I have plenty more un-synthesised evidence with particular emphasis on actin making sense of the senses and as the entropic arrow of time. I hope this book will generate sufficient interest to allow me to go there.

A universal rethink of a wide variety of science puzzles could benefit in light of actin's pivotal role. Placing actin as the central missing piece of the puzzle provides a convincing answer to a host of science puzzles by getting to the point of the root cause and a chance to get back on track with nature with successful strategies:

Disease

The interconnected risks of disease are apparent in an article published on disease patterns through a deep dive into the data of medical records, which unburied previously invisible connections (Chmiel et al., 2014). Researchers hope to use these networks to generate hypotheses about how diseases operate at the molecular level. A greater degree of interconnectivity may become apparent with actin in mind and successful prevention and treatment strategies may be developed by re-focussing attention on actin as root-cause. Actin as root cause of Alzheimer's and other degenerative diseases have been identified (Kurakin and Bredesin, 2015) and in the Actin' Badly section in the Being Human by Design section I propose some other chief suspects.

Migraine

Migraine is characterised by waves (diagnostic of actin) of pain and the associated visual 'aura' (reminded me of the eyes self-organisation/ inherent design through its cytoskeleton, not DNA Kumaramanickavel et al., 2015) consisting of jagged lines (reminded me of actin's barbed ends), emanates from the visual cortex at the back of the brain and spreads forwards- suggested as being caused by dilation and distention of blood vessels (Purdy and Dodick, 2017). Connections between vitamin D, pain and sleep have recently been identified (de Oliviera et al., 2017), with no known mechanism. **I wonder What If!** Is it as simple as actin is over assembling due to breakdown of the treadmilling process (Bugyi and Carlier, 2010) and the barbed ends (Shekhar et al., 2016) are the root cause?

Magnetic Sense of Direction - Migration

Trilobites migrated across the ocean in long orderly queues, reportedly coordinated by chemotaxis (Gramling, 2016). Mechanisms in bird migration are reported as an unsolved mystery (Hamzelou, 2012). My information suggests otherwise: Synchronised migration patterns occur in response to ATP (mitochondria) related environmental triggers (Seebacher and Post, 2015). The role of the eyes and pineal gland with an oscillatory feedback loop in determining the direction have been identified (Cassone and Westneat, 2012), providing a spatial organisation mechanism through triangulation.

I wonder What If! The process could be as simple as being environmental cues entrained in the membrane orchestrated by actin coincidence detection, actin vibration and triangulation relative to the earths electromagnetic field providing direction.

Lost Human Senses

Hints that humans sense the magnetic field have only recently been shown from brain wave activity (Temming, 2019). No wonder with actin as the universal architect! Since the human success story relies so heavily on migration, have we just lost our sense of it since it is no longer important? Could losing our receptiveness to our natural senses be the root cause of our current crises or is that just stretching my imagination too far.

It's not just my imagination Giles Hutchins ideas of (R) EVOLution: Separateness to Connectedness thinks so too https://www.youtube.com/watch?v=t2mUebq5PXU

Photosynthesis

The Earth's magnetic field is undergoing alarming and unprecedented shifts (Hess, 2019) – Digging up fossil fuels can hardly be helping. What If! we understood and capitalised on mastering the mystery of photosynthesis could make the world of difference. A U-tube video by the Royal Society (2016) provides a useful overview of the Quantum Secrets of Photosynthesis, an amazingly efficient energy conversion process. Other publications identify the role of quantum vibration, spatial organisation and polarity (O'Reilly and Olaya Castro, 2014; Roberts, 2015; Choi, 2016; Zhang et al., 2016) missing the point of actin. The complex biochemical interdependence of chloroplasts with mitochondria (Hoefnagel et al., 1998) points in actin's direction. Actin's role in photo-orientation of chloroplasts has been described (Tagaki, 2003) although the mechanism was far from understood.

Synchronicity

Synchronised co-operative behaviour appears to be universal in nature, as a mechanism to achieve full potential (Ellis, 2018). Actin as the key closes the open questions posed as to the root cause (Couzin, 2018). Birds flying in formation also provides other energetic clues, with wing beat synchronised to achieve ultimate cooperation to optimise energy efficiency to achieve an aerodynamic sweet spot (Portugal et al., 2014), supporting actin as the physical form of information.

Viral Seasonality

Emergence of latent herpes simplex viruses in neurons (Brown, 2017) is likely to be attributable to reduced resistance as a consequence of the fitness of the actin cytoskeleton and lack of vitamin D. The 'mystery' of viral seasonality (Fisman, 2012) and 'flu specifically (Moorthy et al., 2012) may also be similarly explained as a consequence (immune reaction) and a coincidence (membrane-actin coincidence detectors, Softky, 1994), taking the actin-hibernation seasonal piece into consideration (Hindle and Martin, 2013; Onufriev et al. 2016).

Actin Really

I also have a rapidly growing list I am calling 'Actin Really' where I have evidence or a hunch that actin is root cause or I wonder Why? they took so long to work it out. It's always easy in hindsight. I plan to pull these together once this book has helped create an appetite for more.

Mixed Cultural Reaction - Realising My Own Potential

I have tested my ideas continuously throughout their development as it felt appropriate. As if by Magic! the right people, the right information and enough resources appeared at the right time in order to encourage me to keep moving and make my journey viable. Just when I thought I was losing my balance, more encouragement arrived in the nick of time, often from unexpected sources.

I've only just realised I needed to be tested with baby steps along the way otherwise I could never have managed to make the quantum leap to writing this book from a standing start without all the dummy runs trying to get my work published as a citizen scientist outside the science establishment. Also very encouraging to know that (R)EVOLutionary science often comes from unexpected outsiders just like me (https://en.wikipedia.org/wiki/The_Structure_of_Scientific_Revolutions).

However fragile my evidence is perceived to be, especially by those it doesn't suit, it is strong enough with the external support and encouragement I have received, to withstand any adverse reactive pressure as a foundation well worth building on.

Open Sesame- Surreal!

Spirituality is something I have only opened my eyes to as a consequence of my own enlightening experience when I attended the Design Indaba conference by pure coincidence on my 'trip' of a lifetime in 2015 (see the Surprise! My Dummy Run Section) – Feeling is Believing. It was priceless and wondered at the time how to bottle it so everyone could share the experience. Encountering open minded pioneers and scientists operating in the US at the edge of science and spirituality arrived very early in the process to maintain my momentum. Had the sequence of events been different, I can't imagine what the consequences might otherwise have been. Special thanks to some key players already engaged in the realisation of what is humanly possible, without whom this would be mission impossible:

Barbara Marx Hubbard: US futurist and driving force behind 'Conscious Evolution', her book that was instrumental in my journey. Barbara spoke at the United Nations Culture of Peace in September 2016, and described 2 missing pieces of the puzzle to the birth of a new humanity. I have the second piece of the puzzle – the process for People to connect their creativity (https://www.facebook.com/watch/?v=10153877344253870). So delighted she found time for a wonderful Skype meeting and an email trail with such encouraging enthusiasm before she disappeared on her Planetary Mission initiative.

Dr. Bruce Lipton: He is an internationally recognised leader in bridging science and spirituality. His bestseller 'The Biology of Belief' was also instrumental. The first of several occasions we have met was at a conference in September 2016 and he continues to show great interest. With any luck he will have agreed to write the Foreword. He described me as an 'out of the box visionary for the biology of consciousness' in an e-mail to Barbara. A beautiful note he wrote in my journal:

'Dear Friend and Evolutionary "Upstart!"! Your quest for a better understanding of evolution is much appreciated. The best insights are from those thinking outside of the box. I LOVE your thinking. Follow your heart and vision- I believe in YOU! I truly believe that your work will profoundly alter our world for the better.'

He warned me of the simplicity of my theory causing an adverse reaction, something he knows too well from his own experience as a stem cell biologist. That could help explain why breaking into the closed science establishment has proven to be so challenging.

Nassim Haramein and William Brown (Resonance Academy). Their quest is to unify science, community and consciousness in resonance with nature and cosmos. A perfect potential fit!

I briefly met Nassim in October 2016 at the premier of his film 'The Connected Universe' which explores new understandings in science that reveal a bigger picture of interconnection than we have ever imagined. This eye opening film explores how the fundamental experience of being human is also about connection… and how this experience may be altered by these breakthroughs in science. http://www.theconnecteduniversefilm.com/. (Coincidentally starts with What If! Everything is connected – exactly what got the ball rolling for me.)

Nassim referred to a science paper about to be published which tackled the science of consciousness- head-on, to include the role of the cytoskeleton as the biological interface with the quantum field. This was already on my radar and music to my ears. The paper The Unified Space Memory Network has since been published (Haramein et al., 2016). I started sharing ideas with William Brown who co-authored the paper, as to my ideas of the biological interface.

William very generously wrote me this endorsement back in April 2017: “Dr Head is at the forefront of consciousness and connectivity science. Bringing together research from numerous fields the work and vision offers a far-ranging synthesis of ideas that, when applied, will shape the course of investigation and development in the science of a truly connected universe.”

A collaborative paper with William, Bruce and I never quite got off the ground. I have developed my ideas quite considerably since then. I wonder What If! Hopefully this excites him enough to get a trip to Hawaii.

Partially Open Science Establishment

I tried to connect with a wide range of scientists who I thought should have been interested, judging by their recent publication history and expression of interest. I received no reply.

I reconnected with practising Professors I know of old in July 2016 with a view to joint publication. The feedback on my first attempt was a double edged sword- not quite as positive as I had hoped but very useful nevertheless. Apparently my work read as a catalogue of key ideas, but did not present a logically developing sequence and that the usual treatment for this type of review is to deal with discoveries in chronological order to the present state of understanding.

Indeed, the artificial separation of nature into a plethora of science information silos, the divide and conquer approach described by Gurel et al., (2014), each of which is growing independently at different rates, has resulted in a lack of connectivity of the holistic picture. This has slowed down the rate of change in knowledge from which we currently suffer and why current science practice has missed the point. Is that not the point!? Another example of how business as usual is not going to crack it and a more open minded approach and mechanism are required.

I carried on regardless, tried again and to my delight they found my ideas very well researched, exciting, very interesting, provocative and should be read, but way, way outside of their comfort zone as a philosophical piece.

Philosophy - Bridging the Science Gap

Philosophy has been identified as the likely source of the answer to formulating new concepts and theories to drive scientific change and progress. A place traditional practising scientists cannot afford to financially go (Laplane et al., 2019).

Modern science without philosophy will run up against a wall, due to information overload complicating interpretation, increasing fragmentation of information silos into more scientific sub disciplines and the emphasis on methods and empirical results will drive training progressively shallower rather than digging into a deeper understanding so vital to the scale of change required. Here's hoping this book will get the right attention to facilitate and accelerate the process.

Citizen Science –People Power Potential Model!

Science culture positively discourages innovation (Lieff Benderley, 2016) and science policy is only now slowly waking up to the realisation that the value of science may lie in engagement of citizen scientists / People Power Potential (Felt, 2016), not just published papers and patents. The nearly universal link between the age of past knowledge and tomorrow's breakthroughs in science and technology provides an alarming picture (Mukherjee et al., 2017). Add to that the wasted time from lack of openness, ridicule of initial breakthrough and lack of transparency of existing knowledge to build on, the true picture is even less attractive. The complexity of the science publication landscape illustrates the point (Boyack et al., 2005) – if only you didn't have to pay £35.94 for the privilege in the closed system we currently operate in.

The financial cost of lack of open science publication was often incredibly frustrating and painful when I was really excited about what I felt I was about to discover- then hitting a paywall like a brick wall! It also had an unexpectedly positive consequence from my point of view. Having to pay was ironically useful as a decision making tool to decide in advance how valuable the content was likely to be in following a thread of evidence or the evaluation of a hunch.

I appear to be materialising as a model (guinea-pig!) for the process. The technology tool I have designed would help facilitate and accelerate it.

Actin's Really Monumental Potential

Monumental

Tuszinsky (2014) predicted actin bridging the gap with the 'Orch OR' theory of consciousness and the potential for monumental breakthroughs. Hopefully my monumental effort will be worth my while.

Just Realised

Actin's potential applications are just starting to be realised and materialise in biotechnology (Kumar and Mansson, 2017) and cytoskeleton computers (Heaven, 2018; Adamatzky et al., 2018) - the original smart material. Tissue engineers have wised up to mechanical tension between tethered cells causing tissue folding (Cepelewicz, 2018b).The health potential of bioelectricity has been recognised in the absence of an understanding of the underlying process (Tyler, 2017). Electroceuticals are just starting to catch on, using electricity to kick- start the immune system (Fox, 2017).

Sustainability: Artificial Photosynthesis Solution. The big question of how quantum energy is conserved at ambient temperature can be answered by actin's physical properties (Huber et al., 2013). Methods are available to investigate the dynamic process (Holweg et al., 2004). Understanding the universal physical process I have realised should help to find successful solutions for clean non-fossil fuel and a more sustainable future.

Filling the Science Blind Spot - Realising Human Experiential Potential

Science is slowly catching up with ancient wisdom and a number of pseudo-scientific and 'spiritual' practices have been progressively accepted- a number of which I believe can be directly attributed to the central role of the actin cystoskeleton as described in the section Being Human by Design (Subsections: Pineal Gland- A Special Case- Science Meet Spirituality and Science Meet More Spirituality- Being Mindful).

Actin as root cause also provides a plausible mechanism for remote information transmission through quantum cognition (Section: Cracking Nature's Code: An Embryonic Potential Physiological Theory of Everything. Subsection: What If! Quantum Effects are Actin's Potential) and opens the door to understanding and acceptance of phenomenological human experiences such as aetheric energy (Degard, 2018) as an example of inclusive cognition (Antinori et al., 2017). Its pivotal role also serves to bridge current gaps in understanding the fluidity of natural inclusion (Rayner, 2018).

We urgently need to fill the science blind spot which excludes human subjective experience and wake up from a delusion of absolute knowledge (Frank et al., 2019) and embrace creating a new scientific culture, in which we see ourselves both as an expression of nature and as a source of nature's self-understanding.in order to realise our full human potential.

An elegant call for action for an open minded approach to consciousness and recognise science currently beyond the imaginative reach of traditional science is made by Cardeña (2014), or indeed beyond technological grasp. In particular the concept of exceptional claims requiring exceptional evidence resonates, hence the heavy weight of evidence presented in the bibliography at the back of this book (just the tip of the iceberg).

Life is Just an Illusion - Time to Get Real

Biomimicry: Technology to Potentially Turn Life Around

Framework Thinking

Framework thinking allows simplicity, standardisation and optimum efficiency of a process, in order to quickly navigate complex problems and arrive at novel solutions. It provides extraordinary opportunity for evaluation and inter-operability between stakeholders. The infinite possibilities of applications make a framework approach essential to streamline a seamless transition from the earliest experience in education and allow customisation depending on application and a scalable operating system.

Success Framework –Simple Solution to Master Complexity

However embryonic or unpopular the theory, I believe it is sufficiently robust as proof of principle to justify biomimicry of the universal evolutionary design principles. These provide a robust foundation of a 'Success Framework' as a creative thinking and decision making tool. Embedded in an emergent 3 layer model to reflect the micro/macro/mega-evolution levels of complexity, it has the potential to realise full personal and collective potential to facilitate and accelerate the cultural change the world is crying out for.

The framework design is integral to a multi-purpose 'People Power' toolkit which includes additional design features. These include mindfulness tools to get you in the right frame of mind and technological tools such as data visualisation for superior pattern recognition in routine business use and less familiar Artificial Intelligence to do the hard work to validate or evaluate your ideas.

This could unlock human intelligence 'People Power' potential in conjunction with AI, potentially the biggest innovation (Johnson, 2016), to facilitate and accelerate innovation on an individual, collective and global scale. A simple system to master complexity and co-create a new and exciting Protopian future in a small window of opportunity to avoid further chaos in order to achieve explosive positive technological exponential growth (Berman et al., 2016). Impossible to imagine but obvious in hindsight and a successful new reality according to nature's architectural design not those of human construction.

Social Network Redesign- Multilayer Model

Design flaws in the internet, recognised by its own inventor (Weaver, 2018) and Facebook (Metz, 2017) need to be overcome, with considerable potential for multilayer models identified in a redesigned social network, to spread positive ideas and behaviour (Falk and Bassett, 2017).

Realising Full Potential of 'People Power'

By virtue of its universal nature, the toolkit has infinite possibilities to facilitate and accelerate a redesign and systemic reconnection with nature and cultural success it can potentially deliver in all sorts of ways:

Individually: As an educational and continuous learning tool, in order to realise full creative potential and self-satisfaction through resonance with a sense of purpose by complementing the natural inherent design process. Development of AI skills as second nature in anticipation of future needs is key to success and a mechanism for determining novelty and value of contribution to support Universal Basic Assets (Gorbis, 2017)- Universal Basic Income is not a viable solution (Rushkoff,2018)with the potential to culturally backfire.

Collectively: An organisational creative mechanism as a foundation for future good work. To encourage, engage and grow creativity, purpose and success. A redesigned social medium architecture founded on self-information not inherited wisdom, to encourage positive action, defuse anti-social behaviour, facilitate a shift to Governance not Government and deliver collective intelligence for positive cultural change.

Globally: A mechanism to create scalable solutions to overcome the 'system immune response' (Conway et al., 2017) by anticipating barriers through process trust and transparency (i.e. Blockchain), establishing opportunity and value through balanced risk assessment, context dependant action plans and avoiding unexpected negative consequences of the immune cascade reaction. A mechanism to act as responsible managers of the natural evolutionary process in a regenerative culture and minimise the pain of transition to a value based economy essential to successfully complement AI.

Reality Check - Create an Internet Social Safety.Net

There is a realisation there are a distinct lack of social safety nets and action is needed in this space (Nair, 2018). My ideas should help to fill the gap:

Safe Thinking Space: Playing it safe to developing innovative ideas is not the solution. The key to privacy of thinking beyond current limits of imagination is afforded by the 3 layered emergent infrastructure of the success framework, with a safe space to develop ideas until you have enough evidence and confidence to share with like minds and encourage 'crazy' ideas to avoid potential silence or ridicule through ignorance and/or arrogance and fear of losing income streams, or much worse if the impact is not favourable to the Powers that be have a vested interested- all conspiring to maintain the *status quo* in the absence of an alternative and the associated anxiety and frustration - Lalochezia springs to mind. Personal protection through anonymity where your work speaks for itself is something I know would be an extremely valuable asset of Blockchain.

Viability: Blockchain has the potential power to create a good viable internet Safety.Net, laying the path for the transformed internet of Value (Wolpert, 2018), the realisation of the full potential of People Power through Universal Basic Assets (Gorbis, 2017), removing the energy blocks and resistance to change of the current system and paving the way for philanthrocapitalism (Scofield, 2018). This is a new idea which involves using private wealth imaginatively, constructively and systematically to attack our fundamental problems and potentially accelerates funding valuable innovation - Less patience required to reduce frustration (Dodgson and Gann, 2018) where AI validates a human hunch, evaluates novelty and impact of human creativity and we get to exercise our imaginations and do the interesting, exciting and valuable stuff. An AI Human synergetic symbiosis: Evolutions innovation process.

Transparency and Trust: Copy right financial fights (Mullin, 2018), inefficient innovation (National Academies of Sciences, Engineering, and Medicine. 2017), legal blocks (Mullin, 2017), would all become a non-sense if AI and Blockchain were engaged to their full potential: Patents would be patently obvious; Legal-ise to Legal-Ease; People Power not expert culture killing the innovator (Krook, 2017).

Unexpected Consequences - Beware the Negative and Accentuate the Positive

Sense of Excitement Turned on its Head Leads to Frustration, Temptation and Evil!!!: The explosive nature of actin (Xu and Robinson, 2013), has amazing potential for positive change. In the absence of the freedom to pursue your passion, it can negatively backfire through frustration (frustrated – action when your path to purpose is blocked, giving the irresistible urge to release potential energy), temptation to take risks to buck the system and when the worst comes to the worst be EVIL (the polar opposite to the meaning of being aLIVE where self-esteem through gratitude leads to wellbeing (Lin, 2015) not letting off a head of steam). Getting to the root cause, breakdown the barriers and releasing the pressure is vital for success in engineering a peaceful transition to the new story of excitement. Otherwise it could be Game Over!

System Change: Successful solutions to current crises will need to be mindful of their interconnectivity in order to avoid unexpected consequences of failure as a consequence of current limitations. Limitations on our current strategy for change have been repeatedly highlighted, with some spectacular misses scored and without learning the lessons in what should be a continuous performance feedback loop. A report from the US identifies how big data analysis (Koebler, 2016; the macroscopic view- equivalent to layer 3) has uncovered previously unconnected catastrophic events, mainly as a consequence of fundamentally flawed Government decisions- What a surprise!.It provides a predictive mechanism for future impact of decisions- a powerful tool for the future. Smarter decision making processes are critical for success as is implicit in my technology design.

Terrifying: 5G Technology: Some of the potential human health hazards of electromagnetic fields are already recognised, all symptomatic with interference of actin, including neuropsychiatric disorders including depression (Pall, 2016), with 5G described as the "Stupidest Idea In The History of The World" (Walia, 2019), driven by the power of money. Bearing in mind the potential interference with actin's universal communication system, the potential impact is mind blowing.

> "Humans beings always do the most intelligent thing…after they've tried every stupid alternative and none of them have worked"- Buckminster Fuller

Happiness: One surprisingly unexpected consequence of a universal system redesign is likely to be of greater happiness, a potentially valuable performance indicator of success if we manage to achieve our full potential : Purposeful valuable work (Burrell, 2017) as opposed to a job, in light of the advent of AI; counteracting a wide variety of diseases and disorders such as depression and anxiety (Baer, 2017) from not feeling on form ; natural highs from great ideas (Jones, 2017) all conspiring in an upward spiral to transform the human condition. The publication of a World Happiness Report (Helliwell et al., 2018) is most encouraging, with close alignment of human happiness with Sustainability Development Goals (SDGs) through collaborative action to design and deliver better lives, rather than income and wealth.

Inter- Operability

A standardised approach to generate ideas to solve puzzles of all descriptions, evaluating the novelty, impact, risk, consequences (unexpectedly negative or surprisingly positive) of ideas with proportionate reward in an inter-operable system with visibility and transparency and a dynamic continuous improvement feedback loop is vital to success.

This would provide a mechanism to create innovation domains (Rotman, 2009) to facilitate and accelerate innovation- where a domain emerges piece by piece from its individual parts and provides a technology toolbox that can be applied across many industries, breaking down the barriers of science silos, ensure we avoid replication of effort, gaps, missing any tricks, unexpected consequences and create bigger and better ideas than otherwise possible and the potential to facilitate and accelerate global transformation, providing a 21st Century solution for success in catalysing cultural change and rebooting globalisation (Chakhoyan, 2017).

Independent Reaction

Royal Society for the encouragement of Arts, Manufacture and Commerce (RSA)

According to their website: 'The RSA's mission is to enrich society through ideas and action. We believe that all human beings have creative capacities that, when understood and supported, can be mobilised to deliver a 21st century enlightenment. Supported by our 29,000 Fellows, we share powerful ideas, carry out cutting-edge research and build networks and opportunities for people to collaborate, helping to create fulfilling lives and a flourishing society. If you're a champion of new ideas and want to support our mission of a 21st century enlightenment, then we want to hear from you: Such a perfect potential fit!

Potential High Social Impact-YES!!!!?

I submitted my idea as a 'Catalyst of Cultural Change' to their UK Challenge for a Citizens Economy- Making Today's Economy Work for Tomorrow in 2017, which was selected and made it through to the final evaluation- How Exciting!

The evaluation panel replied as follows: 'We regret to inform you that your idea did not win the challenge. We hope you will be pleased to hear that our panel sensed that your idea would have high social impact, although they did struggle to understand how it would work as a policy. They felt your idea was more akin to a theory or framework that is trying to better manage something which may be beginning to happen organically at many levels. They thought your idea could be strengthened with further consideration of: Which actors should be involved in drawing up and testing prototypes; Where will funding come from; How it will be publicised and campaigned for to gain public support'.

I had hoped they would pick up the ball and run with it. Not the right organisation or the right time most likely. Most encouraged that they sensed the potential high social impact. New Rules of the Game will require a single framework and inter-operability for a fundamental redesign. Throwing it back to me was really unexpected.

Happy to take one for the team
until they turn up!

Game On- Managing the Transition

Power Games- Nature's Game Rules- People Power to the Rescue

Super to Soft Power

The architecture of global economic power has been fully revealed with disconcerting and unstable super connectivity of a capitalist network that runs the world (Coghlan and MacKenzie, 2011). A 'Soft Power' alternative has real possibilities for the cultural redesign required (MacDonald, 2018).

Powers that Be

The Powers that be are talking the talk - no one is talking the walk, let alone walking it, with everyone left to their own devices in global transformation required in light of AI according to the World Economic Forum (Schwab, 2018) - a recipe for disaster. The United Nations recognise the interconnectivity of issues surrounding conflict and call for a people- centred approach in principle to achieve world peace (Joëlle & Day, 2017), with no idea as to How? Their biggest policy challenge.

Nature's Game People Power Rules – Baby Steps to Cooperation and Reciprocity

Getting back on track with the inherent design laws of nature to create an upward spiral to success is the way forward. If only we could achieve the full infinite potential of our human capital – People Power- and optimise finite global resources with a clear way ahead. What a wonder-fuelled world it could be.

Growing up and waking up to our full protopian future to celebrate individual uniqueness and realise full human potential in a happier cooperative global culture where reciprocity is the name of the game is vital to success (Frank et al., 2018). People Power has a very valuable part to play in saving the planet, not least in the impact of individual baby steps to move more rapidly towards a sustainable safe space (Corner, 2018).

Reboot- People Power Team Players

The World Economic Forum (Chakhoyan, 2017) gives an outline of what is required to reboot the current global system, particularly to provide tools of engagement with the individual as a focus and new ways of working between agencies such as the United Nations (UN).Their biggest policy challenge is how to incorporate People Power in the growth process, with inclusivity as the key to success. Their solutions are still amorphous and the crystallisation of mine as a catalyst of cultural change holds a real possibility as to the solution. Empowerment of the next generation to realise their full potential, otherwise the last generation, is vital to success.

Now is the time to take the power into our own hands and co-create a new story of value. Capitalising on well-being and unrealised human potential by correcting the flaws of capitalism (Butler, 2018) amongst other quick-fix human superficial illusions of happiness and success.

Target Practice Education VIPs

Education Is A Set Up Causing Foul Play

The education system set up for the third industrial revolution has stifled our natural curiosity and imagination. We are living in the fourth industrial revolution and education has a lot of catching up to do. We are currently preparing students for jobs that don't yet exist, using technologies that haven't yet been invented to solve problems we don't even know are problems yet- Brilliant!.

The teenage brain brings its own challenges. Hormones have an interesting impact on encouraging risky behaviour with often disastrous consequences. There is currently a mismatch between the maturity of executive function of the prefrontal cortex, responsible for the risk decision process and the hormone driven risk taking limbic system. Embedding value – guided goal pursuit is recognised as being important in directing positive behaviour (Davidow et al., 2018). Anti-social behaviour is born out of boredom and frustration seeking the excitement that should be our birth right! In the absence of the freedom to pursue your passion, it can negatively backfire through frustration (frustrated – action when your path to purpose is blocked, giving the irresistible urge to release potential energy), temptation to take risks to buck the system and when the worst comes to the worst be EVIL (the polar opposite to the meaning of being aLIVE where self-esteem through gratitude leads to wellbeing, not letting off a head of steam).

Put the FUN Back into Fundamental

Risky teenage excitement seeking behaviour can be taken care of by encouraging development of the fundamental inherent design process from an early age through redesign of education. Harnessing negative passions and redirecting them to a unified vision of a mature human-unity to defuse our immature state of play, refuse the *status quo* and put the FUN back into fundamental.

The solution is recognised as adaptive learning, powered by artificial intelligence - I couldn't agree more. The Global Education Futures Report (https://futuref.org/educationfutures) describes a wisdom based approach to education and societal transformation is right on track with a perfect fit – it's as if I had written the paper myself!

Idea Generation- Baby Steps – Learning to Run

Key design features of the technology toolkit are in order to harness People Power potential and get us up and running in the same direction as safely and as soon as possible:

Sweett Shop: A pick and mix to choose from for positivity to get you in the right frame of mind. (Sweett is a family surname, not a misspelling.)

Direction: Finding your why? Identifying important pieces of your unique personal puzzle to be solved in order to finding purpose and the right direction is fundamental so you can stand out from the crowd. The success framework will help you tease out what excites your IT factor from your biological information software system and determine the consequences of your actions before you decide whether and what action to take.

Goal Setting Action Plan: Adjustment of achievable targets for success will be a highly valuable asset. The science of goal setting explains why: Baby achievable steps towards meaningful goals are best to avoid disappointment.

Prioritisation of what is valuable and worth the return on investment and the willpower to stay motivated is so important.

Look where you are going! Visioning success and identifying baby steps forwards are most likely to be successful, without putting excess pressure on the system and blowing the circuitry of your mind. Identifying resources (information, human, financial), barriers to be removed or planned for and achievable timescales may be the next step, depending on the context and scale of the piece of the puzzle to be solved.

Sweett Dreams: Visibility of the status of all action plans in a dashboard provides self-satisfaction you are on track and in control, designed to help you sleep like a baby. During sleep, information is self-organised so strong connections from brain trains of thought are consolidated into strong lines of evidence as valuable information and weak ones subject to synaptic pruning. Export of poor ideas, decisions and emotionally expensive people and import of the opposite leads to a happy balance and a great start for positively facing the next day.

Team Work- Social Medium for Cultural Growth- High Five!

Global connectivity is key to success. Delivering collective intelligence through social media is the theory but far from effective in practice: Facebook has emerged as our foremost (anti) social media network, built on flimsy foundations for college dating, not without design flaws. It has fallen far short of giving people voice and meaning - indeed it has taken us in the opposite direction by threatening democracy with fake news and MEanness. It is rapidly losing face as a consequence. The need for more alternatives is fully recognised. Social intelligence to (R)EVOLutionise education facilitated by creative educational tools in a connected economic sharing model is on the cards (Martin, 2017) according to the World Economic Forum- another perfect fit.

Following Iceland's lead, orchestrating a new social medium founded on natural highs to best generate ideas founded on self-information, thinking deep not at face value, to avoid the scourge of self-it is (a recognised mental disorder) and fake news, is what I have in mind. Make life more meaningful for yourself and you can become the centre of attention for all the right reasons.

Quick Wins- Pre-formulated Innovation Ideas Brought to Life

The European Responsible Research and Innovation initiative has taken a baby step forwards in the right direction with pre-formulated ideas. If only you knew they were there! People Power engagement is recognised as their missing piece of the puzzle. Prioritisation through the evaluation process and embedding in the 3-layered social medium model has the potential to bring it to life, prove its value and generate some quick wins. There are more than likely to be other initiatives that may be further ahead of the game to build on. Who knows!

Start the Ball Rolling- Philanthrocapitalism

Philanthrocapitalism is a new idea which involves using private wealth imaginatively, constructively and systematically to attack our fundamental problems and potentially accelerate funding valuable innovation (Scofield, 2018). The tragic fire at Notre Dame Cathedral just goes to show how much money is magically available in a crisis. World war has historically shown money to be no object. If you are fortunate enough to have more money than you usefully know what to do with, the solution I propose would save you the effort of trying to decide what you believe to be the best value and return on investment, with confidence and optimum self-satisfaction.

Kick Start- Soft Economic Landing

Safe delivery during a potentially rocky transition period is the biggest challenge, with a parallel social economy within the belly of capitalism proposed to facilitate the transition and bridge the gap between the old and new (R)EVOLutionary game (Bollier, 2017). The Institute for the Future has set out their manifesto and action plan for Universal Basic Assets (Gorbis, 2017) aka People Power as a game changer and most promising path forward.

My ideas potentially provide a ready-made mechanism to kick-start the process.

Hold Back- Economics Running Ahead

Principles of a new care economy, re-imagines value (Bollier, 2017). Parallel thinking on my part and complementary key attributes of the success framework are identified:

- Abstraction as a thinking technique
- Value estimation chain
- Open value network framework
- Money - Shift to purpose driven for human freedom
- Contemporary re-emergence of gift exchange

I LOVE this statement: 'Economics is a tool to build the future we want, not a science to describe the reality we endure.' I came cross it in an online blog which makes a lot of sense to me (https://musingsofayounglondoner.wordpress.com/2019/02/15/the-most-important-and-least-asked-questions-in-economics/). It gives me great hope that we can turn the world around by shifting towards an economy of value. It was also my introduction to the book 'The Value of Everything' (Mazzucato, 2018) and a useful reminder of an economy distributive by design to create global balance in a safe and just sustainable space for humanity (Raworth, 2018) fit for the future, to complete the current picture as I see it - Perfect Fit for Potentially Answering Everything!

Economics has the right idea in developing a new framework. Unfortunately a Cambrian explosion of frameworks are competing with each other (what a Surprise!) and need reigning in. A single framework makes much more sense to me as a standard operating procedure, as a crystallisation of simple evolutionary design rules in order to catalyse and realise full potential through inter-operability and scalability, to deliver the transformation required from human hungry caterpillar to global butterfly (Chase, 2016).

Social Safety. Net

There is a realisation there are a distinct lack of social safety nets and action is needed in this space (Nair, 2018). My ideas include an internet safety.net with viability afforded by Blockchain which should help to fill the gap and a real possibility of facilitating and accelerating the process with minimum pain.

Responsible Managers Required

New Ball - Prototype Development

Having inadvertently picked up the ball in the evolutionary game of life, I would be delighted to pass it on, subject to a satisfactory handover. Better to shape an idea of a ball to kick around than no ball at all.

The success framework developed into a prototype seed AI for standardisation and realisation of a viable internet safety.net is the first vital step- But Who? Unfortunately, but not surprisingly in the current state of affairs, AI has become the new space race with different initiatives internationally with competition and potential conflicts of interest (Gershgorn, 2018). Also, in the absence of a better understanding of how human intelligence works that I bring to the game, we are in grave danger of merging AI with human stupidity and changing the nature of the human race to the bottom (Fahim, 2017).

Human Upper Hand – Synchronicity – Nature's Secret Success Story

In order for humans to keep the upper hand over AI and avoid what might otherwise become an AI Apocalypse to achieve a successful global transformation, controlled management is vital for success. From personal experience, an interesting symptom of tuning in with nature's inherent design is super-sensitivity and frustration as to the futility of our current reality. A global tune in and wake up would have its own hazards, further compounded in the absence of a soft landing. Careful handling in a controlled experiment is required. After all, synchronicity is nature's secret success story to achieving full action potential.

Guiding Hand- Global Governance - United Nations Step Up Or Handover

Cooperation with a global governance body for implementation, including operational system feedback, guidance and continuous improvement before general release without negative explosive unexpected consequences, is what I have in mind.

The United Nations realises that they need to up their game in order to achieve a level global playing field, or risk being cut out or held to ransom by super powers. In offering its services to convene players and build consensus will require a significant organizational and cultural shift (Muggah and Kavanagh, 2018).

Time for United Nations to fully realise their role and engage 'People Power' in the process, their biggest policy challenge for achieving inclusive growth and ensuring AI works for and not against humanity (Muggah and Cavanagh, 2018) or handover.

Soft Power Experiment- UK Great Candidate

Out of the chaos that currently reigns in the UK as a consequence of the political fallout of Brexit, comes the potential for a monumental light at the end of the tunnel. The concept of 'Soft Power' has the real possibility of controlling the experiment and getting the ball rolling to collectively head in the right direction and putting the 'Great' back into Great Britain as the 'Soft Power Super Power' (MacDonald, 2018).

People Power Rules!

Natural or Foregone Conclusion

What If! We Don't Redesign Change Its Really Game Over.

A leading economist warns of World War as an inevitable consequence of disintegrating capitalism (Hannapi, 2019). Time is of the essence. The infrastructure build to evolve the future that the People Power piece of the puzzle fits into makes a case for systemic cultural design that it is already late in the game. For successful delivery, projects have to start in 2020 as case studies, to spread by 2030, in time for 2050 by when all institutions and practices must be based on biomimicry principles in order to successfully design the future (Brewer, 2016).

What If! It's Not My Idea? So What

Mother Nature doesn't put all her eggs in one basket- I know I am not the only one thinking along the same lines/ on the same wavelength, there is too much evidence to ignore. I find myself in a lonely and frustrating space, having spent so much time in perseverance (through-severance – Sincere apology to cutting myself off from friends!) making as much sense as I can before feeling enough confidence to reach out and share the excitement of my ideas with you.

If you have a better idea, I would LOVE to hear it and hope you can resist the temptation of sending unhelpful feedback along the lines of the 'LOVE' letters sent to Richard Dawkins in response to his idea of the 'selfish gene': https://www.facebook.com/The.Dutch.Atheist/videos/888290881345437/?t=33.

Let's Get Cracking and Hold Your Peace

Evolution and the emergence process have separately been identified as the key to future success frameworks across a variety of disciplines including economics and public policy (Wilson and Gowdy, 2013). Maximum efficiency would be provided by their convergence into a unified framework with interoperability and People Power engagement. By co-creating a cultural growth dynamic tipping point we can achieve a global butterfly effect of human metamorphosis to maturity (Chase, 2016) away from the unsustainable hungry caterpillar we currently represent.

Transition from the current operating system to a redesigned one forced by the hand of Artificial Intelligence is likely to be slow and messy (Lindmark, 2017) unless we are smart enough to prepare for it by creating a new system to jump onto- a 21st Century Noah's Ark.

Solutions have been identified in principle but I am not aware of any other pragmatic cards on the table built on such a solid foundation to facilitate and accelerate the process as painlessly as possible.

We are all designed to be Human Kind by nature. If! I can help one person in real life that is my great natural privilege and pleasure. I really hope it resonates and helps you to be inspired to take action of your own in some shape, way or form. With that knowledge would really help me to turn my own life around. For you to have bought this book has already helped my confidence and self-belief to unscramble my brain, come out of my shell, stop beating myself up, not be chicken and get the flying monkeys off my case. What If! This can help humanity as a whole create a new reality of human-unity – we can all live happily ever after.

Time to Get Cracking- Otherwise Game Over!

Now is the right time to connect in order to create a People Power voice for change. 'Nothing is more powerful than an idea whose time has come'- Victor Hugo

I really hope you will lend your support and join forces on www.headcase.global

Surprise! My Dummy Run

I can't imagine anybody being anywhere near as surprised as I am to find myself writing this book, particularly since it appears to be my last resort in spreading such potentially good news!

I have always been passionate about challenging the *status quo.* The example that springs to mind was the history teacher that killed my passion for the subject. We all had the same text book. Lessons involved the teacher dictating from the book and we dutifully followed her lead. Until one bright spark (guess who?) asked the question everyone else was probably thinking- Why? and What If! we agree to read a chapter in advance of the lesson and discuss. Imagine the response. Another one is perhaps more powerful and pertinent in a real 'life' lesson. Discussing belief systems, mine conflicted with the teacher's. The marks for my homework magically plummeted and I was threatened with extra work. I made a pact with a classmate and plagiarised their homework. They got an A, I got a D. I challenged the teacher and my marks went back to normal - Magic!

I studied microbiology at degree level, only because there was more future career promise than biology, my natural attraction. It was such a 'new' subject my biology teachers at school hadn't heard of it. Undeterred, I put in so much effort and did such a great job that much to my surprise and my family's I was offered a PhD with four or five to choose from. The one I selected was in the space organisation of fungi focussing on hexagonal crystals (Head, 1987) offering the first opportunity to do something really different, not just following in others footsteps. In my thesis are secreted some wonderful images of 3D scanning electron micrographs of the intact actin cytoskeleton using a (R)EVOLutionary cryo- technique. They were taken on an exciting day trip to the Natural History Museum since their electron microscope was much more powerful, for the magnification necessary for the required resolution. It felt quite surreal, as if I had been miniaturised and was roaming around the inside of a cell – such a powerful and privileged experience. Even now science is still surprised the process I was investigating is actively managed (Steinberg at al., 2017; by actin would be no surprise to me) and that hexagons are so universal to natural design. I never imagined in a million years it would be so incredibly relevant 30 years later as the secret ingredient to life - Magic!

I have enjoyed a successful, diverse and high profile career in industry on an International platform, predominantly South Africa, specialising in system design and development from scratch, seamlessly filling the gaps between sectors or dis-entangling broken ones to optimise efficiency by embedding them in technological solutions.

Challenging the *status quo* and making change in my professional capacity at a local, national and international level, as well as working under enormous emotional pressure in broken businesses held together by human glue, has been a competitive strength and advantage. I often wondered whether it was really worth the effort. All my experience turns out to have been great practice in my dummy run.

My greatest success should have been my brain-child Expert-Ease that connected industry experts with technology tools to facilitate and accelerate industry innovation- described as the 'Biggest Innovation' at its launch in 2012. For no apparently good reason, the mission was not accomplished. It has since turned out to be a dummy run for maturing into something MUCH bigger. The parallels with this new (R)EVOLutionary path are extra-ordinary!

An (un)fortunate side-effect /unexpected consequence of this was to raise my personal self- belief beyond my natural modesty and level of expectation of life. Also the pain and frustration of financial bullying as an independent consultant became a frequent *modus operandi* with too many trips to register court papers and monies owed materialised like Magic! This is what led me to take a completely different path- apparent failure leading to what I hope to be much greater success on what is currently a labour of LOVE.

Conceived by coincidence: In 2015 I had had enough. My idea was to start a South African arts and crafts import business for fun - something I had promised myself some years before through my many business trips there. I planned to attend the Design Indaba exhibition in Cape Town but pushed the wrong button on the electronic booking form. To my great Surprise! I had booked into the Design Indaba International Design Conference #make change, which turned out to be the 'trip' of a lifetime at the conference itself and all sorts of Magic! synchronicities.

The conference was a highly positively emotionally charged experience, with a host of presentations providing triggers and clues that strongly resonated with me and set my intuition on fire.

To illustrate: Finding your WHY? Or sense of purpose through connecting the human dots of advertising; How purpose drives creativity and change; Peripheral (Design) Thinking; An incredibly powerful recital by Dennis Hopper of Rudyard Kipling's If! (https://www.youtube.com/watch?v=wY3dJYU542Q); Thinking your way out of the Normal; Beauty and lack of inclusivity of Black women.

Last but far from least six degrees of separation Kevin Bacon - the connections and science of which really got the ball rolling. The presenter is my Wizard of Oz, giving me the lion's courage on my yellow brick road. Special thanks to him for his continued encouragement on this new road somewhere over the rainbow.

At the exhibition I sat in the Dreams for Africa Chair- I dreamed of a crystal ball and a magic wand: Enhancing natural predictive powers through data visualisation technology, creating natural highs as do Magic! mushrooms, Magic! synchronicities and potentially the big reveal of the Magic! Key to information theory and natural mysteries - hopefully now not too far removed from making the dream a reality.

Another extra-ordinary coincidence that I should be geographically so close to the birthplace of humanity, and have found myself on a fascinating parallel life path, with the technology tool to potentially give birth to the next phase of evolution itself- No wonder it's a painful process.

Not something I designed myself, it just evolved through natural curiosity about the wonder of life and stretching my imagination as to wondering Why? and How? Life operates and What If! it could be easier and much more wonderful than my own fractured reality.

I was so busy with my head down sniffing out clues and leads, fascinated and excited by the exploratory process, by the time I looked up I could hardly believe how far I had strayed off the beaten track and to my enormous Surprise! I had somehow managed to pick up the ball in the evolutionary 'Game of Life'. Fortunately I have crossed paths with enough people who believed in and encouraged me, just in the nick of time, enough to keep me going and soothe the shell shock and fragility to my system. I would be delighted to pass the ball on to the next responsible manager.

Now is my chance to deliver, it's just a big question of scale- the surprising mathematics of life and civilisation (West,2014). Amazingly I had all the ingredients; I just needed to change my own recipe from the apparent failure of Expert-Ease to MUCH greater success.

> "You can't connect the dots looking forward; you can only connect them looking backwards. So you have to trust that the dots will somehow connect in your future. You have to trust in something – your gut, destiny, life, karma, whatever. Because believing that the dots will connect down the road will give you the confidence to follow your heart even when it leads you off the well worn path; and that will make all the difference."- Steve Jobs

Being naturally modest, this is ridiculously outside my comfort zone and taken a disproportionate amount of effort and time to write this section. Even more ridiculous, as a self-confessed technophobe I have designed a technology tool to potentially turn the world around: The combination of human and artificial intelligence will define humanity's future as the 'Biggest Innovation' (Johnson, 2016). As it turns out I would be Surprised If! we could design life to be any more complicated, less fair, balanced or satisfying for the vast majority than it currently is.

Not Surprisingly! This is all much to the amusement of those who know me. I am hoping to Surprise! them and get the last laugh as the butt of the ultimate cosmic joke and get a well-earned super energy boost in self-confidence to boot.

> "Here's to the crazy ones, the misfits, the rebels, the troublemakers, the round pegs in the square holes... the ones who see things differently — they're not fond of rules... You can quote them, disagree with them, glorify or vilify them, but the only thing you can't do is ignore them because they change things... they push the human race forward, and while some may see them as the crazy ones, we see genius, because the ones who are crazy enough to think that they can change the world, are the ones who do."– Steve Jobs

We are all designed to be Human Kind by nature. If! I can help one person in real life that is my great natural privilege and pleasure- sometimes making me a soft touch for the unscrupulous and criticised by the hard boiled.

I really hope it resonates and helps you to be inspired to take action of your own in some shape, way or form. That knowledge would really help me to turn my own life around. For you to have bought this book has already helped my confidence and self-belief to unscramble my brain, come out of my shell, stop beating myself up, not be chicken and get the flying monkeys off my case.

My most favourite song from an early age is 'I'd Like to Teach the World to Sing' by the New Seekers (https://www.youtube.com/watch?v=ZWKznrEjJK4). What If! I can help humanity as a whole create a new reality of human-unity – we can all live happily ever after.

I really hope you will lend your support and join forces on www.headcase.global

Bibliography

Aanen, D.K. and Egleton, P. (2017). Symbiogenesis : Beyond the endosymbiosis theory? *Journal of Theoretical Biology*, https://doi.org/10.1016/j.jtbi.2017.08.001

Abdill, R. J. & Blekhman, R. (2019). What bioRxiv's first 30,000 preprints reveal about biologists. *Nature,* https://doi.org/10.1101/515643.

Adamatzky, A., Tuszynski, J., Pieper. J., Nicolau, D.V., Rinalndi, R., Sirakoulis, G., Erokhin, V., Schnauss, J. and Smith, D.M. (2018). Towards Cytoskeleton Computers. A proposal. https://arxiv.org/abs/1810.04981

Adami, C. and Hintze, A.(2018). Thermodynamics of evolutionary games. *Physical Review E*, https://doi.org/10.1103/PhysRevE.97.062136

Aerts, D. and de Bianchi, M.S. (2015). The unreasonable success of quantum probability I: Quantum measurements as uniform fluctuations. *Journal of Mathematical Psychology*, https://doi.org/10.1016/j.jmp.2015.01.003

Aftab, O., Cheung, P., Kim, A., Thakkar, S., Yeddanapudi, N. (2001). Information Theory and the Digital Age. Massachusetts Institute of Technology, http://web.mit.edu/6.933/www/Fall2001/Shannon2.pdf

Ahrens, S., Zelenay. S., Sancho, D., Hanč, P., Kjær, S., Feest, C., Fletcher. G., Durkin, C., Postigo, A., Skehel, M., Batista, F., Thompson. B., Way, M., e Sousa,C.R. and Schulz, O.(2012). F-Actin Is an Evolutionarily Conserved Damage-Associated Molecular Pattern Recognized by DNGR-1, a Receptor for Dead Cells. *Journal of Immunity*, https://doi.org/10.1016/j.immuni.2012.03.008

Allen, J.F. (2017). The CoRR theory for genes in organelles. *Journal of Theoretical Biology.* https://doi.org/10.1016/j.jtbi.2017.04.008

Ananthaswamy, A. (2014). Root intelligence: Plants can think, feel and learn. *New Scientist,* https://www.newscientist.com/article/mg22429980-400-root-intelligence-plants-can-think-feel-and-learn/ Roots of Consciousness https://doi.org/10.1016/S0262-4079(14)62338-1

Ananthaswamy, A. (2017). The brain's 7D sandcastles could be the key to consciousness. *New Scientist,* https://www.newscientist.com/article/mg23531450-200-the-brains-7d-sandcastles-could-be-the-key-to-consciousness/; Throwing Shapes, https://doi.org/10.1016/S0262-4079(17)31924-3

Anders, S., de Jong, R., Beck, C., Haynes, J-D. and Ethofer, T. (2016). A neural link between affective understanding and interpersonal attraction. *PNAS,* https://doi.org/10.1073/pnas.1516191113

Antinori, A., Carter, O.L. and Smillie, L.D. (2017). Seeing it both ways: Openness to experience and binocular rivalry suppression. *Journal of Research in Personality,* https://doi.org/10.1016/j.jrp.2017.03.005

Antony, J.W., Schönauer, M., Staresina, B.P. and Cairney, S.A. (2018). Sleep Spindles and Memory Reprocessing. *Trends in Neurosciences,* https://doi.org/10.1016/j.tins.2018.09.012

Araque, A. and Navarette, M. (2010). Glial cells in neuronal network function. *Philosophical Transactions Of The Royal Society B*, https://doi.org/10.1098/rstb.2009.0313

Ashtiani, A. and Azgomi, M.A. (2015). A survey of quantum-like approaches to decision making and cognition. *Mathematical Social Sciences*, https://doi.org/10.1016/j.mathsocsci.2015.02.004

Atasoy, S., Roseman, L., Kaelen, M., Kringelbach, M.L., Deco, G. and Carhart-Harris, R.L. (2017). Connectome-harmonic decomposition of human brain activity reveals dynamical repertoire re-organization under LSD. *Scientific Reports,* https://doi.org/10.1038/s41598-017-17546-0

Aylett, C.H.S., Löwe, J. and Amos, L.A.A. (2011). Chapter one - New Insights into the Mechanisms of Cytomotive Actin and Tubulin Filaments. *International Review of Cell and Molecular Biology,* https://doi.org/10.1016/B978-0-12-386033-0.00001-3

Baarlink, C., Plessner, M., Sherrard, A., Morita, K., Misu, S., Virant, D., Kleinschnitz, E.M., Harniman, R., Alibhai, D., Baumeister, S., Miyamoto, K., Endesfelder, U, Kaidi, A. and Grosse, R. (2017). A transient pool of nuclear F-actin at mitotic exit controls chromatin organization. *Nature Cell Biology*, https://doi.org/10.1038/ncb3641

Baer, D. (2017). Psychologists Think They Found The Purpose Of Depression. https://www.huffpost.com/entry/psychologists-think-they-found-the-purpose-of-depression_b_589e58a8e4b080bf74f03c45. Accessed 5 November 2018.

Bain, N. and Bartolo, D. (2019). Dynamic response and hydrodynamics of polarized crowds. *Science*, https://doi.org/10.1126/science.aat9891

Barbey, A.K. (2017). Network Neuroscience Theory of Human Intelligence. *Trends in Cognitive Sciences*, https://doi.org/10.1016/j.tics.2017.10.001

Barras, C. (2016). You are junk: Why it's not your genes that make you human. https://www.newscientist.com/article/mg23130840-400-the-junk-that-makes-you-human/

Bartók, A.P., De, S., Poelking, C., Bernstein, N., Kermode, J.R., Csányi, G. and Ceriotti, M. (2017). Machine learning unifies the modeling of materials and molecules. *Science Advances,* https://doi.org/ 10.1126/sciadv.1701816

Bekrater-Bodmann, R., Foell, J., Diers, M., Kamping, S., Rance, M., Kirsch, P., Trojan, J., Fuchs, X., Bach, F., Çakmak, H.K., Maaß, H. and Flor, H. (2014). The importance of synchrony and temporal order of visual and tactile input for illusory limb ownership experiences - an FMRI study applying virtual reality. *PLOS ONE*, https://doi.org/10.1371/journal.pone.0087013

Bement, W.M. and von Dassow, G. (2014). Single cell pattern formation and transient cytoskeletal arrays. *Current Opinion in Cell Biology,* https://doi.org/10.1016/j.ceb.2013.09.005

Berdahl, A., Brelsford, C., De Bacco, C., Dumas, M., Ferdinand, V., Grochow, J.A., Hébert-Dufresne, L., Kallus, Y., Kempes, C.P., Kolchinsky, A., Larremore, D.B., Libby, E., Power, E.A., Stern, C.A. and Tracey, B. (2017). Dynamics of beneficial epidemics. *Physics and Society,* arXiv:1604.02096 / https://arxiv.org/abs/1604.02096

Berman, A.E., Dorrier, J. and Hill, D.J.(2016) How to Think Exponentially and Better Predict the Future. Singularity Hub, https://singularityhub.com/2016/04/05/how-to-think-exponentially-and-better-predict-the-future/ Accessed 22 January 2019

Bezanilla, M., Gladfelter, A.S., Kovar, D.R. and Lee, W-L. (2015). Cytoskeletal dynamics: A view from the membrane. *Journal of Cell Biology,* https://doi.org/10.1083/jcb.201502062

Blanquie, O. and Bradke, F. (2018). Cytoskeleton dynamics in axon regeneration. *Current Opinion in Neurobiology,* https://doi.org/10.1016/j.conb.2018.02.024

Bocchio, M., Nabavi, S. and Capogna, M. (2017). Synaptic Plasticity, Engrams, and Network Oscillations in Amygdala Circuits for Storage and Retrieval of Emotional Memories. *Neuron,* https://doi.org/10.1016/j.neuron.2017.03.022

Boldogh, I.R. and Pon, L.A. (2006). Interactions of mitochondria with the actin cytoskeleton. *Biochimica et Biophysica Acta - Molecular Cell Research,* https://doi.org/10.1016/j.bbamcr.2006.02.014

Bollier D. (2017). Re-imagining Value: Insights from the Care Economy, Commons, Cyberspace and Nature. Available at : https://www.boell.de/en/2017/03/07/re-imagining-value-insights-care-economy-commons-cyberspace-and-nature

Bongiorni, F. (2017). To advance science we need to think about the impossible. *New Scientist,* https://www.newscientist.com/article/mg23331153-800-to-advance-science-we-need-to-think-about-the-impossible/ Accessed 12 March 2019

Bower, B. (2018). The economics of climate change and tech innovation win U.S. pair a Nobel. *Science News,* https://www.sciencenews.org/article/economics-climate-change-and-tech-innovation-win-us-pair-nobel?utm_source=email&utm_medium=email&utm_campaign=latest-newsletter-v2

Boyack, K., Klavans, R. and Börner, K. (2005). Mapping the backbone of science. *Scientometrics,*https://doi.org/10.1007/s11192-005-0255-6

Bregman, R.(2017). A growing number of people think their job is useless. Time to rethink the meaning of work. World Economic Forum. https://www.weforum.org/agenda/2017/04/why-its-time-to-rethink-the-meaning-of-work. Accessed 6 November 2018.

Brewer, J. (2016). Culture Design Labs—Evolving the Future. *Medium,* https://medium.com/age-of-awareness/culture-design-labs-evolving-the-future-94455c446ff5 Accessed 12 March 2019

Brooks, M. (2015). Is quantum physics behind your brain's ability to think? *New Scientist* https://www.newscientist.com/article/mg22830500-300-is-quantum-physics-behind-your-brains-ability-to-think/ https://doi.org/10.1016/S0262-4079(15)31805-4 A bit in two minds, https://doi.org/10.1016/S0262-4079(15)31805-4

Brown, G.D. (2012). Actin' dangerously. *Nature,* https://doi.org/10.1038/485589a

Brown, J.C. (2017). Herpes Simplex Virus Latency: The DNA Repair-Centered Pathway. *Advances in Virology,* https://doi.org/10.1155/2017/7028194

Bugyi, B. and Carlier, M-F. (2010). Control of Actin Filament Treadmilling in Cell Motility. *Annual Review of Biophysics,* https://doi.org/10.1146/annurev-biophys-051309-103849

Burrell, T. (2017). A meaning to life: How a sense of purpose can keep you healthy. *New Scientist,* https://www.newscientist.com/article/mg23331100-500-a-meaning-to-life-how-a-sense-of-purpose-can-keep-you-healthy/ Why am I here? https://doi.org/10.1016/S0262-4079(17)30179-3

Butler,S. (2018). The Impact of Advanced Capitalism on Well-being: an Evidence-Informed Model. *Human Arenas,* https://doi.org/10.1007/s42087-018-0034-6

Byczkowicza, N., Ritzau-Josta, A., Delvendahl, I. and Hallermanna, S. (2018). How to maintain active zone integrity during high-frequency transmission. *Neuroscience Research,* https://doi.org/10.1016/j.neures.2017.10.013

Cardeña, E. (2014). A call for an open, informed study of all aspects of consciousness. *Frontiers in Human Neuroscience,* https://doi.org/10.3389/fnhum.2014.00017

Cassone, V.M. and Westneat, D.F. (2012) The bird of time: cognition and the avian biological clock. *Frontiers in Molecular Neuroscience,* https://doi/10.3389/fnmol.2012.00032

Cedernaes, J., Schönke, M., Orzechowski- Westholm, J., Mi, J., Chibalin, A., Voisin, S., Osler, M., Vogel, H., Hörnaeus, K., Dickson, S.L., Bergström-Lind., S., Bergquist, J., Schiöth, H.B., Zierath, J.R., and Benedict, C. (2018). Acute sleep loss results in tissue-specific alterations in genome-wide DNA methylation state and metabolic fuel utilization in humans. *Science Advances,* https://doi/10.1126/sciadv.aar8590

Cepelewicz, J. (2017a). Life's First Molecule Was Protein, Not RNA, New Model Suggests. *Quanta Magazine,* https://www.quantamagazine.org/lifes-first-molecule-was-protein-not-rna-new-model-suggests-20171102/ Accessed 6 November 2018

Cepelewicz, J. (2017b). The End of the RNA World Is Near, Biochemists Argue. *Quanta Magazine,* https://www.quantamagazine.org/the-end-of-the-rna-world-is-near-biochemists-argue-20171219/ Accessed 6 November 2018

Cepelewicz, J. (2018a). Cell by Cell, Scientists Map the Genetic Steps as Eggs Become Animals. *Quanta Magazine,* https://www.quantamagazine.org/cell-by-cell-scientists-map-the-genetic-steps-as-eggs-become-animals-20180426/ Accessed 6 November 2018

Cepelewicz, J. (2018b). Tissue Engineers Hack Life's Code for 3-D Folded Shapes. *Quanta Magazine,* https://www.quantamagazine.org/tissue-engineers-hack-lifes-code-for-3-d-folded-shapes-20180125/ Accessed 6 November 2018

Chaabra, E.S. and Higgs, H.N. (2007). The many faces of actin: matching assembly factors with cellular structures. *Nature Cell Biology*, https://doi.org/10.1038/ncb1007-1110

Chakhoyan, A. (2017). 3 ways to reboot globalization. *World Economic Forum*, https://www.weforum.org/agenda/2017/02/3-ways-to-reboot-globalization/ Accessed 16 January 2019

Chamorro- Premuzic,T.(2014). Curiosity Is as Important as Intelligence. *Harvard Business Review,* https://hbr.org/2014/08/curiosity-is-as-important-as-intelligence Accessed 12 March 2019

Chandel, N.S. (2015). Evolution of Mitochondria as Signaling Organelles. *Cell Metabolism,* https://doi.org/10.1016/j.cmet.2015.05.013

Chase, C. (2016). The Global Butterfly Effect. *Dreamcatcher Reality,* http://dreamcatcherreality.com/global-butterfly-effect/ Accessed 13 March 2019

Chen, L., Shi,K., Frary, C.E., Ditzel, N., Hu, H., Qiu, W. and Kassem, M. (2015). Inhibiting actin depolymerization enhances osteoblast differentiation and bone formation in human stromal stem cells. *Stem Cell Research,* https://doi.org/10.1016/j.scr.2015.06.009

Chen, J., Chang Leong, Y., Honey, C.J., Yong, C.H., Norman, K.A. and Hasson, U. (2017a). Shared memories reveal shared structure in neural activity across individuals. *Nature Neuroscience,* https://doi.org/10.1038/nn.4450

Chen, C., Liu, S., Shi, W-q., Chaté, H. and Wu, Y. (2017b). Weak synchronization and large-scale collective oscillation in dense bacterial suspensions. *Nature,* https://doi.org/10.1038/nature20817

Chen, L., Hu, H., Qiu, W., Shi, K. and Kassem, M. (2018). Actin depolymerization enhances adipogenic differentiation in human stromal stem cells. *Stem Cell Research,* https://doi.org/10.1016/j.scr.2018.03.010

Chittka, L. and Wilson,C. (2018). Bee-brained. *Aeon,* https://aeon.co/essays/inside-the-mind-of-a-bee-is-a-hive-of-sensory-activity. Accessed 12 January 2019

Chmiel, A., Klimek, P. and Thurner, S. (2014). Spreading of diseases through comorbidity networks across life and gender. *New Journal of Physics,* https://doi.org/10.1088/1367-2630/16/11/115013

Choi, C.Q. (2016). Scientists Capture 'Spooky Action' In Photosynthesis. *Inside Science,* https://www.insidescience.org/news/scientists-capture-spooky-action-photosynthesis Accessed 6 November 2018

Christophel, T.B., Christiaan Klink, P., Spitzer, B., Roelfsema, P.R., Haynes, J-D. (2017). The Distributed Nature of Working Memory. *Trends in Cognitive Sciences,* https://doi.org/10.1016/j.tics.2016.12.007

Church, D. (2018). *Mind to Matter: The Astonishing Science of How Your Brain Creates Material Reality.* Hay House Inc.. ISBN-10: 1401955258; ISBN-13: 978-1401955250

Claessen, D., Rozen, D.E., Kuipers, O.P., Søgaard-Andersen,L. and van Wezel, G.P. (2014). Bacterial solutions to multicellularity: a tale of biofilms, filaments and fruiting bodies. *Nature Reviews Microbiology*, https://doi.org/10.1038/nrmicro3178

Clarke, D., Morley, E. and Robert, D. (2017). The bee, the flower, and the electric field: electric ecology and aerial electroreception. *Journal of Comparative Physiology A*, https://doi.org/10.1007/s00359-017-1176-6

Clynes, T. (2017). How to raise a genius: lessons from a 45-year study of super-smart children. *Nature*, https://doi.org/10.1038/537152a

Coghlan, A. (2017). Psychedelic drugs push the brain to a state never seen before. *New Scientist*, https://www.newscientist.com/article/2128192-psychedelic-drugs-push-the-brain-to-a-state-never-seen-before/ LSD puts brain in state we've not seen before https://doi.org/10.1016/S0262-4079(17)30815-1

Coghlan, A. (2018). We've cracked the brain's emotion code and it may help depression. *New Scientist*, https://www.newscientist.com/article/mg23931954-300-weve-cracked-the-brains-emotion-code-and-it-may-help-depression/ Brain's emotion code is cracked https://doi.org/10.1016/S0262-4079(18)31642-7

Coghlan A. and MacKenzie, D. (2011). Revealed – the capitalist network that runs the world. *New Scientist*, https://www.newscientist.com/article/mg21228354-500-revealed-the-capitalist-network-that-runs-the-world/

Cohen, M.X. (2017). Where Does EEG Come From and What Does It Mean? *Trends in Neurosciences*, https://doi.org/10.1016/j.tins.2017.02.004

Colin, A., Bonnemay, L., Gayrard, C., Gautier, J. and Gueroui, Z. (2016). Triggering signaling pathways using F-actin self-organization. *Scientific Reports*, https://doi.org/10.1038/srep34657

Conway, R., Masters, J. and Thorold, J. (2017). *From Design Thinking to Systems Change*. Royal Society of Arts, https://www.thersa.org/globalassets/pdfs/reports/rsa_from-design-thinking-to-system-change-report.pdf ISBN 978-1-911532-03-3

Cooper, E.A., Garlick, J., Featherstone, E., Voon, V., Singer, T., Critchley, H.D. and Harrison, N.A. (2014). You Turn Me Cold: Evidence for Temperature Contagion. *PLOS ONE,* https://doi.org/10.1371/journal.pone.0116126

Cooper, R. L., Thieryl,. P., Fletcher,. G., Delbarre, D.J., Rasch, L.R. and Fraser, G.J. (2018). An ancient Turing-like patterning mechanism regulates skin denticle development in sharks. *Science Advances,* https://doi.org/10.1126/sciadv.aau5484

Corner, A. (2018). Tiny individual decisions really could help avert climate chaos. *New Scientist,* https://www.newscientist.com/article/2158432-tiny-individual-decisions-really-could-help-avert-climate-chaos/ Accessed 9 April 2019

Coulthard, L.G., Hawksworth, O.A. and Woodruff, T.M. (2018). Complement: The Emerging Architect of the Developing Brain. *Trends in Neurosciences,* https://doi.org/10.1016/j.tins.2018.03.009

Couzin, I.D. (2018). Synchronization: The Key to Effective Communication in Animal Collectives. *Trends in Cognitive Sciences,* https://doi.org/10.1016/j.tics.2018.08.001

Crivelli, C. and Fredlund, A.J. (2018). Facial Displays Are Tools for Social Influence. *Trends in Cognitive Sciences,* https://doi.org/10.1016/j.tics.2018.02.006

Cudmore, S., Reckmann, I. and Way,M. (1997). Viral manipulations of the actin cytoskeleton. *Trends in Microbiology,* https://doi.org/10.1016/S0966-842X(97)01011-1

Dahl, M. (2016). Has Hypnosis Finally Been Vindicated by Neuroscience? *The Cut,* https://www.thecut.com/2016/11/has-hypnosis-finally-been-vindicated-by-neuroscience.html? Accessed 1 April 2019

Dance, A. (2016). Inner Workings: Uncovering the neuron's internal skeleton. *PNAS,* https://doi.org/10.1073/pnas.1617651113

Davidow, J.Y., Insel, C. and Somerville, L.H. (2018). Adolescent Development of Value-Guided Goal Pursuit. *Trends in Cognitive Sciences,* https://doi.org/10.1016/j.tics.2018.05.003

Davies, P. (2019). Life's secret ingredient: A radical theory of what makes things alive. https://www.newscientist.com/article/mg24132150-100-lifes-secret-ingredient-a-radical-theory-of-what-makes-things-alive/ Accessed 2 April 2019

Dayel, M.J., Alegadoa, R.A., Fairclough, S.R., Levin, T.C., Nichols, S.A. McDonald, K. and King, N. (2011). Cell differentiation and morphogenesis in the colony-forming choanoflagellate *Salpingoeca rosetta. Developmental Biology,* https://doi.org/10.1016/j.ydbio.2011.06.003

de Lange, F.P., Heilbron, M. and Kok, P. (2018). How Do Expectations Shape Perception? *Trends in Cognitive Sciences,* https://doi.org/10.1016/j.tics.2018.06.002

De Loof, A. (2016). The cell's self-generated "electrome": The biophysical essence of the immaterial dimension of Life? *Communicative & Integrative Biology,* https://doi.org/10.1080/19420889.2016.1197446

De Loof, A. (2017). The evolution of "Life": A Metadarwinian integrative approach. *Communicative & Integrative Biology, http://dx.doi.org/10.1080/19420889.2017.1301335*

De Ninno, A. and Pregnolato, M. (2017). Electromagnetic homeostasis and the role of low-amplitude electromagnetic fields on life organization. *Electromagnetic Biology and Medicine,* https://doi.org/10.1080/15368378.2016.1194293

De Oliveira, D.L., Hirotsu, C., Tufik, S. and Levy Andersen, M. (2017). The interfaces between vitamin D, sleep and pain. *Journal of Endocrinology,* https://doi.org/10.1530/JOE-16-0514

De Pittà, M.D., Brunela, N and Volterra, A. (2016). Astrocytes: Orchestrating synaptic plasticity?*Neuroscience,* https://doi.org/10.1016/j.neuroscience.2015.04.001

de Ramon Francàs, G., Zuñiga, N.R. and Stoeckli, E.T. (2017). The spinal cord shows the way – How axons navigate intermediate targets. *Developmental Biology,* https://doi.org/10.1016/j.ydbio.2016.12.002

De Souza, W. (2012). Prokaryotic cells: structural organisation of the cytoskeleton and organelles. *Memórias do Instituto Oswaldo Cruz,* http://dx.doi.org/10.1590/S0074-02762012000300001

de Waal, F.B.M. (2008). Putting the Altruism Back into Altruism: The Evolution of Empathy. *Annual Review of Psychology,* http://dx.doi.org/10.1146/annurev.psych.59.103006.093625

del Mar Quiroga, M., Morris, A.P. and Krekelberg, B. (2016). Adaptation without Plasticity. *Cell Reports,* https://doi.org/10.1016/j.celrep.2016.08.089

Degard (2018). Aetheric Energy: the Intangible Source of Human Celebrity? *Human Arenas,* https://doi.org/10.1007/s42087-018-0021-y

Dener, E., Kacelnik, A. and Shemesh, H. (2016). Pea Plants Show Risk Sensitivity. *Current Biology,* https://doi.org/10.1016/j.cub.2016.05.008

Diamant, E. (2008). Unveiling the mystery of visual information processing in human brain. *Brain Research,* https://doi.org/10.1016/j.brainres.2008.05.017

Dijksterhuis, A. & Strick, M. (2016). A Case for Thinking Without Consciousness. *Perspectives on Psychological Science,* https://doi.org/10.1177/1745691615615317

Dodgson, M. and Gann, D. (2018). The missing ingredient in innovation: patience. *World Economic Forum,* https://www.weforum.org/agenda/2018/04/patient-capital/ Accessed 22 January 2019

Doeller, C.F., Barry, C. and Burgess, N. (2010). Evidence for grid cells in a human memory network. *Nature,* https://doi.org/10.1038/nature08704

Doherty, G.J. and McMahon, H.T. (2008). Mediation, Modulation, and Consequences of Membrane-Cytoskeleton Interactions. *Annual Review of Biophysics,* https://doi.org/10.1146/annurev.biophys.37.032807.125912

Dolgin, E. (2019). How secret conversations inside cells are transforming biology. *Nature*, http://doi.org/10.1038/d41586-019-00792-9

Dominguez, R. and Holmes, K.C. (2011). Actin Structure and Function. *Annual Review of Biophysics*, https://doi.org/10.1146/annurev-biophys-042910-155359

Dor, D. (2015). *The Instruction of Imagination.* Oxford : Oxford University Press. ISBN: 9780190256623

Dowd, M. (2017). Elon Musk's Billion-Dollar Crusade to Stop the A.I. Apocalypse. *Vanity Fair*, https://www.vanityfair.com/news/2017/03/elon-musk-billion-dollar-crusade-to-stop-ai-space-x Accessed 22 January 2019

Dragoš, A. and Kovács, A.T. (2017). The Peculiar Functions of the Bacterial Extracellular Matrix. *Trends in Microbiology*, https://doi.org/10.1016/j.tim.2016.12.010

Dumontheil, I. (2014). Development of abstract thinking during childhood and adolescence: The role of rostrolateral prefrontal cortex. https://doi.org/10.1016/j.dcn.2014.07.009

Duque, J., Greenhouse, I., Labruna, L. and Ivry, R.B. (2017). Physiological Markers of Motor Inhibition during Human Behavior. *Trends in Neurosciences*, https://doi.org/10.1016/j.tins.2017.02.006

Eisenberg-Bord, M., Shai, N., Schuldiner, M. and Bohnert, M. (2016). A Tether Is a Tether Is a Tether: Tethering at Membrane Contact Sites. *Developmental Cell*, https://doi.org/10.1016/j.devcel.2016.10.022

Eisenstein, M. (2017). A measure of molecular muscle. *Nature*, https://doi.org/10.1038/544255a

Eisenstein, M. (2018). Transparent tissues bring cells into focus for microscopy. *Nature*, https://doi.org/10.1038/d41586-018-07593-6

Ellis,J. (2018). Shawn Achor on the Secret to Reaching Big Potential. *Success*, https://www.success.com/shawn-achor-on-the-secret-to-reaching-big-potential/ Accessed 8 November 2018

Encode (2012). An integrated encyclopedia of DNA elements in the human genome. *Nature,* https://doi.org/10.1038/nature11247

Engen, H.G. and Anderson, M.C. (2018). Memory Control: A Universal Mechanism of Emotion Regulation. *Trends in Cognitive Sciences,*https://doi.org/10.1016/j.tics.2018.07.015

Erez, Z., Steinberger-Levy, I., Shamir, M., Doron, S., Stokar-Avihail, A., Peleg, Y., Melamed, S., Leavitt, A., Savidor, A., Albeck, S., Amitai, G. and Sorek, R. (2017). Communication between viruses guides lysis–lysogeny decisions. *Nature,* https://doi.org/10.1038/nature21049

Fahim, C. (2017). When artificial intelligence meets human stupidity. *World Economic Forum,* https://www.weforum.org/agenda/2017/08/ai-and-human-stupidity/Accessed 12 March 2019

Faivre, N., Filevich, E., Solovey, G., Kühn, S. and Blanke, O. (2017). Behavioural, modeling, and electrophysiological evidence for supramodality in human metacognition. *Journal of Neuroscience,* https://doi.org/10.1523/JNEUROSCI.0322-17.2017

Falk, E.B. and Bassett, D.S. (2017). Brain and Social Networks: Universal Building Blocks of Human Experience. *Trends in Cognitive Sciences,* https://doi.org/10.1016/j.tics.2017.06.009

Felsenberg, J., Jacob, P.F., Walker, T., Jefferis, G.S.X.E., Bock, D.D., Waddell, S., et al. (2018). Integration of Parallel Opposing Memories Underlies Memory Extinction. *Cell,* https://doi.org/10.1016/j.cell.2018.08.021

Felt, U. (2016). Associating citizens with the scientific process from the start. *EuroScientist,* https://www.euroscientist.com/public-engagement-in-research/ Accessed 8 November 2018

Field, C.M. and Lénárt, P. (2011). Bulk Cytoplasmic Actin and Its Functions in Meiosis and Mitosis. *Current Biology,* https://doi.org/10.1016/j.cub.2011.07.043

Fields, R.D. (2016). Learning When No One Is Watching. *Scientific American Mind,* https://doi.org/10.1038/scientificamericanmind0916-56; https://www.nature.com/scientificamericanmind/journal/v27/n5/box/scientificamericanmind0916-56_BX1.html

Fisman, D. (2012). Seasonality of viral infections: mechanisms and unknowns. *Clinical Microbiology and Infection,* https://doi.org/10.1111/j.1469-0691.2012.03968.x

Foley, J.F. (2016). More roles for mitochondria in the immune response. *Science Signalling,* https://doi.org/10.1126/scisignal.aai9319

Forterre P, Filée J, Myllykallio H. (2000-2013). *Origin and Evolution of DNA and DNA Replication Machineries.* In: Madame Curie Bioscience Database [Internet]. Austin (TX): Landes Bioscience. Available from: https://www.ncbi.nlm.nih.gov/books/NBK6360/

Fox, D. (2017). The shock tactics set to shake up immunology. *Nature,* https://doi.org/10.1038/545020a

Frank. M.R., Obradovich, N., Sun, L., Woon, W.L., LeVeck, B.L. and Rahwan, I. (2018). Detecting reciprocity at a global scale. *Science Advances,* https://doi.org/10.1126/sciadv.aao5348

Frank, A., Gleiser, M. and Thompson, E. (2019). The blind spot. *Aeon,* https://aeon.co/essays/the-blind-spot-of-science-is-the-neglect-of-lived-experience? Accessed 22 January 2019

Frum, T. and Ralston, A. (2018). AttrActin' Attention to Early Mouse Development. *Cell,* https://doi.org/10.1016/j.cell.2018.03.078

Fuller, R.B. (2004). *Guinea Pig B: The 56 Year Experiment.* Critical Path Publishing ISBN-10: 097406050X; ISBN-13: 978-0974060507

Fuller, R.B. (2017). *Operating Manual for Spaceship Earth.* Zurich, Switzerland: Lars Muller Publishers. ISBN-10: 0935754016; ISBN-13: 978-3037781265

Galkin, V.E., Orlova A, and Egelman, E.H. (2012). Actin filaments as tension sensors. *Current Biology*, https://doi.org/10.1016/j.cub.2011.12.010

Gardiner, J., McGee, P., Overall, R. et al. (2008). Are histones, tubulin, and actin derived from a common ancestral protein? *Protoplasma*, https://doi.org/10.1007/s00709-008-0305-z

Genon, S., Reid, A., Langner, R., Amunts, K. and Eickhoff, S.B. (2018). How to Characterize the Function of a Brain Region. *Trends in Cognitive Sciences,* https://doi.org/10.1016/j.tics.2018.01.010

Gershgorn, D. (2018). AI is the new space race. Here's what the biggest countries are doing. *Quartz,* https://qz.com/1264673/ai-is-the-new-space-race-heres-what-the-biggest-countries-are-doing/ Accessed 12 March 2019

Geoghegan, J.L., Duchêne, S. and Holmes, E.C. (2017). Comparative analysis estimates the relative frequencies of co-divergence and cross-species transmission within viral families. *PLOS PATHOGENS*, https://doi.org/10.1371/journal.ppat.1006215

Gintis, H. (2007). A framework for the unification of the behavioral sciences. Behavioral and Brain Sciences, https://doi.org/10.1017/S0140525X07000581

Goff, P. (2018). Is the Universe a conscious mind? *Aeon*, https://aeon.co/essays/cosmopsychism-explains-why-the-universe-is-fine-tuned-for-life? Accessed 9 November 2018.

Gorbis,M. (2017). There Could Be a Real Solution to Our Broken Economy. It's Called 'Universal Basic Assets.' *Medium*, https://medium.com/institute-for-the-future/universal-basic-assets-abb08ca2f0fc. Accessed 9 November 2018.

Gordon, D.M. (2016). The Collective Wisdom of Ants. *Scientific American,* https://doi.org/10.1038/scientificamerican0216-44

Gorochowski, T.E., Grierson, C.S. and di Bernardo,M. (2018). Organization of feed-forward loop motifs reveals architectural principles in natural and engineered networks. *Science Advances,* https://doi.org/10.1126/sciadv.aap9751

Gramling, C. (2016). What queues up better than the British? Trilobites! *Science*, https://doi.org/10.1126/science.aah7228

Gruters, K.G., Murphy, D.L.K., Jenson, C.D., Smith, D.W., Shera, C.A. and Groh, J.M. (2018). The eardrums move when the eyes move: A multisensory effect on the mechanics of hearing. *PNAS*, https://doi.org/10.1073/pnas.1717948115

Gurel, P.S., Hatch, A.L. and Higgs, H.N. (2014). Connecting the Cytoskeleton to the Endoplasmic Reticulum and Golgi. *Current Biology*, https://doi.org/10.1016/j.cub.2014.05.033

Hameroff, S. and Penrose, R. (2014). Consciousness in the universe: a review of the 'Orch OR' theory. *Physics of Life Reviews,* https://doi.org/10.1016/j.plrev.2013.08.002

Hammerschlag, R., Marx, B.L. and Aickin, M. (2014). Nontouch Biofield Therapy: A Systematic Review of Human Randomized Controlled Trials Reporting Use of Only Nonphysical Contact Treatment. *The Journal of Alternative and Complementary Medicine* https://doi.org/10.1089/acm.2014.0017 Infographic Available at: https://www.chi.is/infographic/

Hamzelou, J. (2012). Mystery of bird navigation system still unsolved. *New Scientist,* https://www.newscientist.com/article/dn21688-mystery-of-bird-navigation-system-still-unsolved/

Hanappi, G. (2019). From Integrated Capitalism to Disintegrating Capitalism. Scenarios of a Third World War. *Munich Personal RePEc Archive,* URI: https://mpra.ub.uni-muenchen.de/id/eprint/91397

Haramein, N., Brown, W. and Val Baker, A.K.F. (2016). The Unified Spacememory Network: from cosmogenesis to consciousness. *Journal of Neuroquantology*. http://dx.doi.org/10.14704/nq.2016.14.4.961

Harold, F.M. (2005). Molecules into Cells: Specifying Spatial Architecture. *Microbiology and Molecular Biology Reviews,* https://doi.org/10.1128/MMBR.69.4.544-564.2005

Harper, D. (2016). The Pineal Gland: A Modulator Of Human Consciousness? *Dream Catcher Reality*, http://dreamcatcherreality.com/pineal-gland-human-consciousness/ Accessed 10 November 2018

He, S. and Davis, W.L. (2001). Filling-in at the natural blind spot contributes to binocular rivalry. *Vision Research,* https://doi.org/10.1016/S0042-6989(00)00315-1

Head, J.B. (1987). *The Isolation and Characterisation of Hexagonal Crystals of Neurospora crassa and Woronin bodies of Penicillium chrysogenum.* Ph.D. Thesis, University of London.

Heaven, D. (2018). Tiny supercomputers could be made from the skeleton inside your cells. *New Scientist,* https://www.newscientist.com/article/2183165-tiny-supercomputers-could-be-made-from-the-skeleton-inside-your-cells/ Computers made from cell skeletons, https://doi.org/10.1016/S0262-4079(18)31962-6

Helliwell, J., Layard, R., & Sachs, J. (2018). World Happiness Report 2018, New York: Sustainable Development Solutions Network. ISBN 978-0-9968513-6-7 http://worldhappiness.report/ed/2018/

Hernández-Hernández, V., Rueda, D., Caballero, L., Alvarez-Buylla, E.R. and Mariana Benítez, M. (2014). Mechanical forces as information: an integrated approach to plant and animal development. *Frontiers in Plant Science,* https://dx.doi.org/10.3389/fpls.2014.00265

Hess, P. (2019). Unexplained Shifts in Earth's Magnetic Field Imperil Global Navigation. *Inverse,* https://www.inverse.com/article/52392-is-the-earths-magnetic-field-going-to-reverse-in-2018? Accessed 1 April 2019

Hill, P.S.M. (2015). Animal Communication: He's Giving Me Good Vibrations. *Current Biology,* https://doi.org/10.1016/j.cub.2015.09.004

Hindle, A.G. and Martin, S.L. (2013). Cytoskeletal Regulation Dominates Temperature-Sensitive Proteomic Changes of Hibernation in Forebrain of 13-Lined Ground Squirrels. *PLOS ONE,* https://doi.org/10.1371/journal.pone.0071627

Hoefnagel, M.H.N., Atkin, O.K. and Wiskich, J.T. (1998). Interdependence between chloroplasts and mitochondria in the light and the dark. *Biochimica et Biophysica Acta (BBA) – Bioenergetics,* https://doi.org/10.1016/S0005-2728(98)00126-1

Holmes, B. (2017). Why be conscious: The improbable origins of our unique mind. *New Scientist,* https://www.newscientist.com/article/mg23431250-300-why-be-conscious-the-improbable-origins-of-our-unique-mind/ Why be conscious? https://doi.org/10.1016/S0262-4079(17)30930-2

Holweg, C., Süßlin, C. and Nick, P. (2004). Capturing in vivo Dynamics of the Actin Cytoskeleton Stimulated by Auxin or Light. *Plant & Cell Physiology,* https://doi.org/10.1093/pcp/pch102

Houser, K. (2018). Experts Answer: Who Is Actually Going to Suffer From Automation? *Futurism,* https://futurism.com/experts-automation-jobs/ Accessed 22 January 2019

Howes, L. (2018) Why You Should Follow Your Curiosity. *Success,* https://www.success.com/why-you-should-follow-your-curiosity/ Accessed 12 March 2019

Huber, F., Schnauß, J., Rönicke, S., Rauch, P., Müller, K., Fütterer, C. and Käs, J. (2013). Emergent complexity of the cytoskeleton: from single filaments to tissue. *Advances in Physics,* https://doi.org/10.1080/00018732.2013.771509

Ibbotson, P. and Tomasello, M. (2016). What's universal grammar? Evidence rebuts Chomsky's theory of language learning. *Scientific American.* http://cogsys.sites.olt.ubc.ca/files/2016/09/Whats-universal-grammar-Eviden-ce-rebuts-Chomsky%E2%80%99s-theory-of-la.pdf

Ingber, D.E. (2000). The origin of cellular life. *BioEssays,* https://doi.org/10.1002/1521-1878(200012)22:12<1160::AID-BIES14>3.0.CO;2-5

Ingber, D.E. (2003). Tensegrity II. How structural networks influence cellular information processing networks. *Journal of Cell Science,* https://doi.org/10.1242/jcs.00360

Iñiguez, L.P. and Hernández, G. (2017). The Evolutionary Relationship between Alternative Splicing and Gene Duplication. *Frontiers in Genetics*, https://doi.org/10.3389/fgene.2017.00014

Inzlicht, M., Shenhav, A. and Olivola, C.Y. (2018). The Effort Paradox: Effort Is Both Costly and Valued. *Trends in Cognitive Sciences,* https://doi.org/10.1016/j.tics.2018.01.007

Irobalieva, R.N., Fogg, J.M. Catanese Jr, D.J., Sutthibutpong, T., Chen, M., Barker, A.K., Ludtke, S.J., Harris, S.A., Schmid, M.F., Chiu, W. and Zechiedrich, L. (2015). Structural diversity of supercoiled DNA. *Nature Communications,* https://doi.org/10.1038/ncomms9440

James, L.S. and Sakata, J.T. (2017). Learning Biases Underlie "Universals" in Avian Vocal Sequencing. *Current Biology*, https://doi.org/10.1016/j.cub.2017.10.019

Janeway, C.A. Jr, Travers, P., Walport, M., et al. (2001). The complement system and innate immunity. *Immunobiology: The Immune System in Health and Disease.* 5th edition. New York: Garland Science

Jayashankar, V. and Rafelski, S.M. (2014). Integrating mitochondrial organization and dynamics with cellular architecture. *Current Opinion in Cell Biology*, https://doi.org/10.1016/j.ceb.2013.09.002

Jégou, A. and Romet-Lemonne, G. (2016). Single Filaments to Reveal the Multiple Flavors of Actin. *Biophysical Journal*, https://doi.org/10.1016/j.bpj.2016.04.025

Jemth, P., Karlsson, E., Vögeli, B., Guzovsky, B., Andersson, E., Hultqvist, G., Dogan, J., Güntert, P., Riek, R. and Chi, C.N. (2018). Structure and dynamics conspire in the evolution of affinity between intrinsically disordered proteins. *Science Advances,* https://doi.org/10.1126/sciadv.aau4130

Jeong, H., Tombor, B., Albert, R., Oltvai, Z.N. and Barabási, A.-L. (2000). The large-scale organization of metabolic networks. *Nature*, https://doi.org/10.1038/35036627

Jockusch, B.M. and Graumann, P.L. (2011). The long journey: actin on the road to pro- and eukaryotic cells. In: Amara S. et al. (eds) Reviews of Physiology, Biochemistry and Pharmacology 161. *Reviews of Physiology, Biochemistry and Pharmacology*, vol 161. Springer, Berlin, Heidelberg. https://doi.org/10.1007/112_2011_1

Joëlle, J. and Day,A.(2017). Can the UN restore international peace? Maybe, but only from the ground up. *World Economic Forum*, https://www.weforum.org/agenda/2017/02/the-united-nations-can-restore-international-peace-but-first-it-must-reform/ Accessed 12 March 2019

Johnson, B. (2016). The combination of human and artificial intelligence will define humanity's future. *TechCrunch*, https://techcrunch.com/2016/10/12/the-combination-of-human-and-artificial-intelligence-will-define-humanitys-future/ Accessed 10 November 2018

Johnson, B. (2018). The most incredible technology you've never seen. *Medium*, https://medium.com/future-literacy/the-most-incredible-technology-youve-never-seen-1284ae628eec

Jones, B. (2017). Neurons Use a Type Of "Adrenaline Rush" to Communicate With the Immune System. *Futurism*, https://futurism.com/neurons-use-a-type-of-adrenaline-rush-to-communicate-with-the-immune-system/. Accessed 10 November 2018

Kaiser, M. (2017). Mechanisms of Connectome Development. *Trends in Cognitive Sciences*, https://doi.org/10.1016/j.tics.2017.05.010

Kaplan, R., Schuck, N.W. and Doeller, C.F. (2017). The Role of Mental Maps in Decision-Making. *Trends in Neurosciences*, https://doi.org/10.1016/j.tins.2017.03.002

Karmarker, U.R. and Buonomano, D.V.(2007). Telling time in the absence of clocks. *Neuron*, https://dx.doi.org/10.1016/j.neuron.2007.01.006

Katsnelsen, A. (2017). The Shape-Shifting Army Inside Your Cells. *Quanta Magazine*, https://www.quantamagazine.org/how-disordered-proteins-are-upending-molecular-biology-20170118, Accessed 13 November 2018

Kaur, S., Fielding, A.B., Gassner, G., Carter, N.J. and Royle, S.J. (2014). An unmet actin requirement explains the mitotic inhibition of clathrin-mediated endocytosis. *eLife*, https://doi.org/10.7554/eLife.00829

Keeling, P.J. and McCutcheon, J.P. (2017). Endosymbiosis: The feeling is not mutual. *Journal of Theoretical Biology*, https://doi.org/10.1016/j.jtbi.2017.06.008

Kershner,K. (2013). "Is the DNA between genes really junk?" *HowStuffWorks.com*, https://science.howstuffworks.com/life/genetic/dna-between-genes-junk1.htm, Accessed 9 November 2018

Keywell, B. (2017). The Fourth Industrial Revolution is about empowering people, not the rise of the machines. World Economic Forum, https://www.weforum.org/agenda/2017/06/the-fourth-industrial-revolution-is-about-people-not-just-machines Accessed 22 January 2019

Klein, A. (2017). Creative people physically see and process the world differently. *New Scientist* https://www.newscientist.com/article/mg23431222-000-creative-people-physically-see-and-process-the-world-differently/ Creative people perceive more, https://doi.org/10.1016/S0262-4079(17)30767-4

Koebler, J. (2016). Society Is Too Complicated to Have a President, Complex Mathematics Suggest. *MOTHERBOARD*, https://motherboard.vice.com/en_us/article/wnxbm5/society-is-too-complicated-to-have-a-president-complex-mathematics-suggest Accessed 6 March 2019

Kral, N., Ougolnikova, A.H. Sena, G. (2016). Externally imposed electric field enhances plant root tip regeneration. *Regeneration (Oxf)*, https://doi.org/10.1002/reg2.59

Kreitz, C., Schnuerch, R., Gibbons, H. and Memmert, D. (2015). Some See It, Some Don't: Exploring the Relation between Inattentional Blindness and Personality Factors. *PLOS ONE*, https://doi.org/10.1371/journal.pone.0128158

Krook, J. (2017). Expert culture has killed the innovator in workplaces. *The Conversation,* https://theconversation.com/expert-culture-has-killed-the-innovator-in-workplaces-77681 Accessed 22 Janury 2019

Ktena, S.I., Arslan, S., Parisot, S. and Rueckert, D. (2017). Exploring Heritability of Functional Brain Networks with Inexact Graph Matching. https://arxiv.org/abs/1703.10062

Kuhn, J.R. and Pollard, T.D. (2005). Real-Time Measurements of Actin Filament Polymerization by Total Internal Reflection Fluorescence Microscopy. *Biophysical Journal,* https://doi.org/10.1529/biophysj.104.047399

Kumar, S. and Mansson, A. (2017). Covalent and non-covalent chemical engineering of actin for biotechnological applications. *Biotechnology Advances,* https://doi.org/10.1016/j.biotechadv.2017.08.002

Kumaramanickavel, G., Denton, M.J. and Legge, M. (2015). No evidence for a genetic blueprint: The case of the "complex" mammalian photoreceptor. *Indian Journal of Opthalmology,* https://doi.org/10.4103/0301-4738.158093

Kumsta, C., Chang, J.T., Schmalz, J. and Hansen, M. (2017). Hormetic heat stress and HSF-1 induce autophagy to improve survival and proteostasis in *C. elegans. Nature Communications,* https://doi.org/10.1038/ncomms14337

Kurakin, A. and Bredesin, D.E. (2015). Dynamic self-guiding analysis of Alzheimer's disease. *Oncotarget,* https://doi.org/10.18632/oncotarget.4221

Lacal, I. and Ventura, R. (2018).Epigenetic Inheritance: Concepts, Mechanisms and Perspectives. *Frontiers in Molecular Neuroscience,* https://doi.org/10.3389/fnmol.2018.00292

Lane, N. (2017). Serial endosymbiosis or singular event at the origin of eukaryotes? *Journal of Theoretical Biology,* https://doi.org/10.1016/j.jtbi.2017.04.031

Laplane, L., Mantovani, P., Adolphs, R., Chang, H., Mantovani, A., McFall-Ngai, M., Rovelli, C., Sober, E. and Pradeu,T. (2019). Opinion: Why science needs philosophy. *PNAS,* https://doi.org/10.1073/pnas.1900357116

Latham, J. (2017). Genetics Is Giving Way to a New Science of Life. *Independent Science News,* https://www.independentsciencenews.org/health/genetics-is-giving-way-to-a-new-science-of-life/ Accessed 14 November 2018

Leach,J. (2018). 'Give-up-itis' revisited: Neuropathology of *extremis. Medical Hypotheses,* https://doi.org/10.1016/j.mehy.2018.08.009

Ledford, H. (2017). Five big mysteries about CRISPR's origins. *Nature,* https://doi.org/10.1038/541280a

Lenne, P-F. and Trivedi, V. (2018). Tissue 'melting' sculpts embryo. *Nature,* https://doi.org/10.1038/d41586-018-06108-7

Le Page, M. (2016). Speargun-toting superbugs end up shooting each other. *New Scientist,* https://www.newscientist.com/article/2076172-speargun-toting-superbugs-end-up-shooting-each-other/

Levin, M. and Martyniuk, C.J. (2018). The bioelectric code: An ancient computational medium for dynamic control of growth and form. *Biosystems,* https://doi.org/10.1016/j.biosystems.2017.08.009

Lichinchi, G., Zhao, B.S., Wu, Y., Lu, Z., Qin, Y., He, C. and Rana, T.M. (2016). Dynamics of Human and Viral RNA Methylation during Zika Virus Infection. *Cell Host & Microbe,* https://doi.org/10.1016/j.chom.2016.10.002

Lieberman, M.D. (2018). Birds of a Feather Synchronize Together. *Trends in Cognitive Sciences,* https://doi.org/10.1016/j.tics.2018.03.001

Liebeskind, B.J., Hofmann, H.A., Hillis, D.M. and Zakon, H.H. (2017). Evolution of Animal Neural Systems. *Annual Review of Ecology, Evolution, and Systematics,* https://doi.org/10.1146/annurev-ecolsys-110316-023048

Lieff Benderley, B. (2016). How scientific culture discourages new ideas. *Science,* https://doi.org/10.1126/science.caredit.a1600102

Lin, C-C. (2015). Self-esteem mediates the relationship between dispositional gratitude and well-being. *Personality and Individual Differences,* https://doi.org/10.1016/j.paid.2015.04.045

Lindmark, R. (2017). Co-evolving the Phase Shift to GameB by Founding The Ethereum Commons. *Medium*, https://medium.com/@RhysLindmark/co-evolving-the-phase-shift-to-cryptocapitalism-by-founding-the-ethereum-commons-co-op-f4771e5f0c83 Accessed 13 March 2019

Loreto V., Servedio V.D.P., Strogatz S.H., Tria F. (2016). Dynamics on Expanding Spaces: Modeling the Emergence of Novelties. In: Degli Esposti M., Altmann E., Pachet F. (eds) *Creativity and Universality in Language*. Lecture Notes in Morphogenesis. Springer, Cham, https://doi.org/10.1007/978-3-319-24403-7_5

Lovejoy, T.E. and Hannah, L. (2018). Avoiding the climate failsafe point. *Science Advances,* https://doi.org/10.1126/sciadv.aau9981

Lu, S., Ge, G. & Qi, Y. (2004). Ha-VP39 binding to actin and the influence of F-actin on assembly of progeny virions. *Archives of Virology*, https://doi.org/10.1007/s00705-004-0361-4

Lu, Y. (2017). Cell-free synthetic biology: Engineering in an open world. *Synthetic and Systems Biotechnology*, https://doi.org/10.1016/j.synbio.2017.02.003

Luczak, A., McNaughton, B.L. and Harris, K.D. (2015). Packet-based communication in the cortex. *Nature Reviews Neuroscience,* https://doi.org/10.1038/nrn4026

MacDonald, A. (2018). *Soft power superpowers: Global trends in cultural engagement and influence.,* British Council. ISBN: 978-0-86355-920-4. Available at: https://www.britishcouncil.org/sites/default/files/j119_thought_leadership_global_trends_in_soft_power_web.pdf

Maly, I.V. and Borisy, G.G. (2001). Self-organization of a propulsive actin network as an evolutionary process. *PNAS,* https://doi.org/10.1073/pnas.181338798

Marek, M., Merten, O-H., Galibert, L., Vlak, J.M. and van Oers, M.M. (2011). Baculovirus VP80 Protein and the F-Actin Cytoskeleton Interact and Connect the Viral Replication Factory with the Nuclear Periphery. *Journal of Virology*, https://doi.org/10.1128/JVI.00035-11

Maritzen, T. and Haucke, V. (2018). Coupling of exocytosis and endocytosis at the presynaptic active zone. *Neuroscience Research*, https://doi.org/10.1016/j.neures.2017.09.013

Markley, O.W. and Harman, W.W. (1982). *Changing Images of Man (Systems Science and World Order).* Pergamon Press Oxford, New York, Toronto, Sydney, Paris, Frankfurt. ISBN 0-08-024314-2 Hard cover ISBN 0-08-024313-4 Flexicover

Martin, N. (2017). Social intelligence will revolutionize education. Here's how. *World Economic Forum.* https://www.weforum.org/agenda/2017/05/joining-the-dots-in-education-aid/ Accessed 14 April 2019

Maslin, M.A., Shultz, S. and Trauth, M.H. (2015). A synthesis of the theories and concepts of early human evolution. *Philosophical Transactions Of The Royal Society B*, https://doi.org/10.1098/rstb.2014.0064

Matcovitch-Natan, O., Winter, D.R., Giladi, A., Aguilar, S.V., Spinrad, A., Sarrazin, S., Ben-Yehuda, H., David, E., González, F.Z., Perrin, P., Keren-Shaul, H., Gury, M., Lara-Astaiso, D., Thaiss, C.A., Cohen, M., Halpern, K.B., Baruch, K., Deczkowska, A., Lorenzo-Vivas, E., Itzkovitz, S., Elinav, E., Sieweke, M.H., Schwartz, M. and Amit, I. (2016). Microglia development follows a stepwise program to regulate brain homeostasis. *Science,* https://doi.org/10.1126/science.aad8670

Mattson, M.P. (2014). Superior pattern processing is the essence of the evolved human brain. *Frontiers in Neuroscience,* https://doi.org/10.3389/fnins.2014.00265

Maxmen, M. (2017). Machine learning predicts the look of stem cells. *Nature,* https://doi.org/10.1038/nature.2017.21769

Mayo, J.P. and Smith, M.A. (2016). Neuronal Adaptation: Tired Neurons or Wired Networks? *Trends in Neurosciences,* https://doi.org/10.1016/j.tins.2016.12.001

Mazzucato, M. (2018). *The Value of Everything: Making and Taking in the Global Economy.* Allen Lane. ISBN: 9780241188811

Metz, C. (2017). Mark Zuckerberg's answer to a world divided by Facebook is more Facebook. *WIRED,* https://www.wired.com/2017/02/mark-zuckerbergs-answer-world-divided-facebook-facebook/ Accessed 22 January 2019

Miller, M.B. and Bassler, B.L. (2001). Quorum Sensing in Bacteria. *Annual Review of Microbiology,* https://doi.org/10.1146/annurev.micro.55.1.165

Miller, L., Balodis, I.M., McClintock,. H., Xu, J., Lacadie, C.M., Sinha, R. and Potenza, M.N. (2018). Neural Correlates of Personalized Spiritual Experiences. *Cerebral Cortex,* https://doi.org/10.1093/cercor/bhy102

Mitra, P. (2017). Is Neuroscience Limited by Tools or Ideas? *Scientific American,* https://www.scientificamerican.com/article/is-neuroscience-limited-by-tools-or-ideas/

Mogessie, B. and Shuh, M. (2014). Nuclear Envelope Breakdown: Actin' Quick to Tear Down the Wall. *Current Biology,* https://doi.org/10.1016/j.cub.2014.05.059

Moorthy, M., Castronovo, D., Abraham, A., Bhattacharyya, S., Gradus, S., Gorski, J., Naumov, Y.N., Fefferman, N.H. and Naumova, E.N. (2012). Deviations in influenza seasonality: odd coincidence or obscure consequence? *Clinical Microbiology and Infection,* https://doi.org/10.1111/j.1469-0691.2012.03959.x

Morell, V. (2016). Plants can gamble, according to study. *Science,* https://doi.org/10.1126/science.aaf5823

Muckli, L. and Petro,L.S. (2017) The Significance of Memory in Sensory Cortex. *Trends in Neurosciences,* https://doi.org/10.1016/j.tins.2017.03.004

Muggah, R. and Kavanagh, C. (2018). 6 ways to ensure AI and new tech works for – not against – humanity. *World Economic Forum*, https://www.weforum.org/agenda/2018/07/united-nations-artificial-intelligence-social-good/ Accessed 21 January 2019

Mukherjee, S., Romero, D.M., Jones, B. and Uzzi, B. (2017). The nearly universal link between the age of past knowledge and tomorrow's breakthroughs in science and technology: The hotspot. Science Advances, https://doi.org/10.1126/sciadv.1601315

Mullin, E. (2017). 5 Biotech Products U.S. Regulators Aren't Ready For. *MIT Technology Review*, https://www.technologyreview.com/s/603870/5-biotech-products-us-regulators-arent-ready-for/?set=603886 Accessed 22 January 2019

Mullin, E. (2018). Billions of dollars are at stake, so the fight over who owns CRISPR is back in court. *MIT Technology Review*, https://www.technologyreview.com/the-download/611053/billions-of-dollars-are-at-stake-so-the-fight-over-who-owns-crispr-is-back-in/ Accessed 22 January 2019

Murakami,K., Yasunaga, T., Noguchi, T.Q.P., Gomibuchi, Y., Ngo, K.X., Uyeda, T.Q.P. and Wakabayashi, |T. (2010). Structural Basis for Actin Assembly, Activation of ATP Hydrolysis, and Delayed Phosphate Release. *Cell*, https://doi.org/10.1016/j.cell.2010.09.034

Nadappuram, B.P. et al. (2018). Nanoscale tweezers for single-cell biopsies. *Nature Nanotechnology*. https://doi.org/10.1038/s41565-018-0315-8

Nair, L. (2018). In the future of work it's jobs, not people, that will become redundant. *World Economic Forum*, https://www.weforum.org/agenda/2018/10/future-of-jobs-humans-skills-leena-nair/ Accessed 14 November 2018

Nanterme, P. (2017). The real value of the Fourth Industrial Revolution? The benefit to society. *World Economic Forum*, https://www.weforum.org/agenda/2017/01/the-real-value-of-the-fourth-industrial-revolution-the-benefit-to-society Accessed 22 January 2019

National Academies of Sciences, Engineering, and Medicine. (2017). *Advancing Concepts and Models for Measuring Innovation: Proceedings of a Workshop*. Washington, DC: The National Academies Press. https://doi.org/10.17226/23640.

National Academies of Sciences, Engineering, and Medicine. 2018. *Learning Through Citizen Science: Enhancing Opportunities by Design*. Washington, DC: The National Academies Press. https://doi.org/10.17226/25183.

Neuheimer, A.B., MacKenzie, B.R. and Payne, M.R. (2018). Temperature-dependent adaptation allows fish to meet their food across their species' range. *Science Advances,* https://doi.org/10.1126/sciadv.aar4349

Newsome, T.P. and Marzook, N.B. (2015). Viruses that ride on the coat-tails of actin nucleation. *Seminars in Cell & Developmental Biology,* https://doi.org/10.1016/j.semcdb.2015.10.008

O'Bleness, M., Searles, V.B., Varki, A., Gagneux, P. and Sikela, J.M. (2012). Evolution of genetic and genomic features unique to the human lineage. *Nature Reviews Genetics,* https://doi.org/10.1038/nrg3336

O'Reilly, E.J. and Olaya-Castro, A. (2014). Non-classicality of the molecular vibrations assisting exciton energy transfer at room temperature. *Nature Communications,* https://doi.org/10.1038/ncomms4012

Ohkawa, T. and Volkman, V.E. (1999). Nuclear F-actin is required for AcMNPV nucleocapsid morphogenesis. *Virology,* https://doi.org/10.1006/viro.1999.0008

Ohlen, A. (2018). Is the Interstitium Really a New Organ? *The Scientist,* https://www.the-scientist.com/daily-news/is-the-interstitium-really-a-new-organ-29893 Accessed 1 April 2019

Omotade, O.F., Pollitta, S.L. and Zheng, J.Q. (2017). Actin-based growth cone motility and guidance. *Molecular and Cellular Neuroscience,* https://doi.org/10.1016/j.mcn.2017.03.001

Onufriev, M.V., Semenova, T.P., Volkova, E.P., et al. (2016). Seasonal changes in actin and Cdk5 expression in different brain regions of the Yakut ground squirrel (*Spermophilus undulatus*). *Neurochemical Journal,* https://doi.org/10.1134/S1819712416020070

Ouellette, J. (2018). Brains May Teeter Near Their Tipping Point. *Quanta Magazine,* https://www.quantamagazine.org/brains-may-teeter-near-their-tipping-point-20180614/ Accessed 14 November 2018

Pacis, A., Tailleux, L., Morin, A.M., Lambourne, J., MacIsaac, J.L., Yotova, V., Dumaine, A., Danckaert,A., Luca, F., Grenier, J.C., et al. (2015). Bacterial infection remodels the DNA methylation landscape of human dendritic cells. *Genome Research,* https://doi.org/10.1101/gr.192005.115

Pall, M.L. (2016). Microwave frequency electromagnetic fields (EMFs) produce widespread neuropsychiatric effects including depression. *Journal of Chemical Neuroanatomy,* https://doi.org/10.1016/j.jchemneu.2015.08.001

Panko, B. (2016). Carp undergo 'reverse evolution' to get their scales back. *Science,* https://doi.org/10.1126/science.aah7219

Paolicelli, R.., Bolasco, G., Pagani, F., Maggi, L., Scianni, M., Panzanelli, P., Giustetto, M., Ferreira, T.A., Guiducci, E., Dumas, L., Ragozzino, D. and Gross, C.T. (2011). Synaptic Pruning by Microglia Is Necessary for Normal Brain Development. *Science,* https://doi.org/10.1126/science.1202529

Parkinson, C., Kleinbaum, A.M. and Wheatley, T. (2018). Similar neural responses predict friendship. *Nature,* https://doi.org/10.1038/s41467-017-02722-7

Patai, E.Z. and Spiers, H.J. (2017). Cracking the mnemonic code. *Nature Neuroscience,* https://doi.org/10.1038/nn.4466

Patalano, R. (2018). From the Cradle to Society: "As-If" Thinking as a Matrix of Creativity. *Human Arenas,* https://doi.org/10.1007/s42087-018-0032-8

Paulson, S. (2017). Roger Penrose On Why Consciousness Does Not Compute. *Nautilus,* http://nautil.us/issue/47/consciousness/roger-penrose-on-why-consciousness-does-not-compute Accessed 14 November 2018

Pelechowicz, S. (2016). What Yoga Has In Common With Anti-Anxiety Meds. *mindbodygreen,* https://www.mindbodygreen.com/0-28062/what-yoga-has-in-common-with-antianxiety-meds.html Accessed 1 April 2019

Pennici, E. (2016). Peaceful ant-plant partnerships lead to genomic arms races. *Science,* https://doi.org/10.1126/science.aah7229

Perez-Moreno, M. and Fuchs, E. (2006). Catenins: Keeping Cells from Getting Their Signals Crossed. *Developmental Cell,* https://doi.org/10.1016/j.devcel.2006.10.010

Pitsalidis, C., Ferro, M.P., Iandolo, D., Tzounis, L. Inal, S. and Owens, R.M. (2018). Transistor in a tube: A route to three-dimensional bioelectronics. *Science Advances,* https://doi.org/10.1126/sciadv.aat4253

PonteCosta, R., Padamsey, Z., D'Amour, J.A., Emptage, N.J., Froemke, R.C. and Vogels, T.P. (2017). Synaptic Transmission Optimization Predicts Expression Loci of Long-Term Plasticity. *Neuron,* https://doi.org/10.1016/j.neuron.2017.09.021

Popat, R., Cornforth, D.M., McNally, L. and Brown, S.P. (2015). Collective sensing and collective responses in quorum-sensing bacteria. *Royal Society Interface,* https://doi.org/10.1098/rsif.2014.0882

Popkin, G. (2017a). Swirling Bacteria Linked to the Physics of Phase Transitions. *Quanta Magazine,* https://www.quantamagazine.org/swirling-bacteria-linked-to-the-physics-of-phase-transitions-20170504

Popkin, G. (2017b). Bacteria Use Brainlike Bursts of Electricity to Communicate. *Quanta Magazine,* https://www.quantamagazine.org/bacteria-use-brainlike-bursts-of-electricity-to-communicate-20170905/

Portugal, S.J., Hubel, T.Y., Fritz, J., Heese, S., Trobe, D., Voelkl, B., Hailes, S., Wilson, A.M. and Usherwood, J.R. (2014). Upwash exploitation and downwash avoidance by flap phasing in ibis formation flight. *Nature,* https://doi.org/10.1038/nature12939

Pradeu, T. and Vivier, E. (2016). The discontinuity theory of immunity. *Science Immunology,* https://doi.org/10.1126/sciimmunol.aag0479

Prindle, A., Liu, J., Asally, M., Ly, S., Garcia-Ojalvo, J. and Süel, G.M. (2015). Ion channels enable electrical communication in bacterial communities. *Nature,* https://doi.org/10.1038/nature15709

Prinz, W.A. (2014). Bridging the gap: Membrane contact sites in signaling, metabolism, and organelle dynamics. *Journal of Cell Biology,* https://doi.org/10.1083/jcb.201401126

Purdy, R.A. and Dodick, D.W. (2017). Can Anything Stop My Migraine? *Scientific American Mind,* https://doi.org/10.1038/scientificamericanmind0517-36

Rand, D.G., Greene, J.D. and Nowak, M.A. (2012). Spontaneous giving and calculated greed. *Nature,* https://doi.org/10.1038/nature11467

Raworth, K. (2018). *Doughnut Economics: Seven Ways to Think Like a 21st-Century Economist.* Random House Business. ISBN-10: 1847941397; ISBN-13: 978-1847941398

Rayner, A. (2018). The Vitality of the Intangible: Crossing the Threshold from Abstract Materialism to Natural Reality. *Human Arenas,* https://doi.org/10.1007/s42087-018-0003-0

Reimann, M.W., Nolte, M., Scolamiero, M., Turner, K., Perin, R., Chindemi, G., Dłotko, P., Levi, R., Hess, K. and Markram, H. (2017). Cliques of Neurons Bound into Cavities Provide a Missing Link between Structure and Function. *Frontiers in Computational Neuroscience,* https://doi.org/10.3389/fncom.2017.00048

Remis, J.P., Wei, D., Gorur, A., Zemla, M., Haraga, J., Allen, S., Witkowska, H.E., Costerton, W., Berleman, J.E. and Auer, M. (2014). Bacterial Social Networks: Structure and composition of *Myxococcus xanthus* outer membrane vesicle chains. *Environmental Microbiology,* https://doi.org/10.1111/1462-2920.12187

Roberts, J. (2015). Vibrating molecules help plants make use of quantum effects. *Horizon Magazine,* https://horizon-magazine.eu/article/vibrating-molecules-help-plants-make-use-quantum-effects.html# Accessed 5 December 2018

Rokas, A. and Carroll, S.B. (2006). Bushes in the Tree of Life. *PLOS Biology,* https://doi.org/10.1371/journal.pbio.0040352

Romet-Lemonne, G. and Jégou, A. (2013). Mechanotransduction down to individual actin filaments. *European Journal of Cell Biology,* https://doi.org/10.1016/j.ejcb.2013.10.011

Rotman, D. (2009). Shoveling Water. Why does it take so long to commercialize new technologies? *MIT Technology Review,* https://www.technologyreview.com/s/416773/shoveling-water/ Accessed 5 December 2018

Rushkoff, D. (2018). Universal Basic Income Is Silicon Valley's Latest Scam. *Medium,* https://medium.com/s/free-money/universal-basic-income-is-silicon-valleys-latest-scam-fd3e130b69a0 Accessed 5 December 2018

Sætra, H.S. (2018). The Ghost in the Machine. *Human Arenas,* https://doi.org/10.1007/s42087-018-0039-1

Sanford-Burnham Prebys Medical Discovery Institute (2017). "What doesn't kill you makes you stronger: Research identifies cellular recycling process linked to beneficial effects of enduring mild stress." *Science Daily,* https://www.sciencedaily.com/releases/2017/02/170215084050.htm

Sanjuán, R. (2017). Collective Infectious Units in Viruses. *Trends in Microbiology,* https://doi.org/10.1016/j.tim.2017.02.003

Sarchet, P. (2016). Life may have emerged not once, but many times on Earth. *New Scientist,* https://www.newscientist.com/article/mg23130870-200-life-evolves-so-easily-that-it-started-not-once-but-many-times/ Life, spontaneously, https://doi.org/10.1016/S0262-4079(16)31516-0

Sarchet, P. (2018).The epic hunt for the place on Earth where life started. *New Scientist,* https://www.newscientist.com/article/mg23831820-400-the-epic-hunt-for-the-place-on-earth-where-life-started/

Life's true cradle, https://doi.org/10.1016/S0262-4079(18)31066-2

Schluter, J., Schoech, A.P., Foster, K.R. and Mitri,S. (2016). The Evolution of Quorum Sensing as a Mechanism to Infer Kinship. *PLOS Computational Biology,* https://doi.org/10.1371/journal.pcbi.1004848

Schönauer, M., Alizadeh, S., Jamalabadi, H., Abraham, A., Pawlizki, A. and Gais, S. (2017). Decoding material-specific memory reprocessing during sleep in humans. *Nature Communications,* https://doi.org/10.1038/ncomms15404

Schuermann, D., Weber, A.E. and Schär, P. (2016). Active DNA demethylation by DNA repair: Facts and uncertainties. *DNA Repair,* https://doi.org/10.1016/j.dnarep.2016.05.013

Schwab,K.(2018). *Shaping the Fourth Industrial Revolution.* World Economic Forum, ISBN: 9781944835149

Scofield, R. (2018). Philanthrocapitalism and the future of giving. *Raconteur,* https://www.raconteur.net/finance/philanthrocapitalism-future-giving?

Seebacher, F. and Post, E. (2015). Climate change impacts on animal migration. *Climate Change Responses,* https://doi.org/10.1186/s40665-015-0013-9

Seibert, M. (2018). Systems Thinking and How It Can Help Build a Sustainable World: A Beginning Conversation. *Solutions,* https://www.thesolutionsjournal.com/article/systems-thinking-can-help-build-sustainable-world-beginning-conversation/

Sen, B., Xie, Z., Uzer, G., Thompson, W.R., Styner, M., Wu, X. and Rubin, J. (2015). Intranuclear Actin Regulates Osteogenesis. *Stem Cells*, https://doi.org/10.1002/stem.2090

Sezgin, E., Levental, I., Mayor, S. and Eggeling, C. (2017). The mystery of membrane organization: composition, regulation and physiological relevance of lipid rafts. *Nature Reviews Molecular Cell Biology*, https://doi.org/10.1038/nrm.2017.16

Shekhar, S., Pernier, J. and Carlier, M-F. (2016). Regulators of actin filament barbed ends at a glance. *Journal of Cell Science*, https://doi.org/10.1242/jcs.179994

Shermer, M. (2018). Utopia is a dangerous ideal. We should aim for "protopia". *Skeptic Magazine*, https://qz.com/1243042/utopia-is-a-dangerous-ideal-we-should-aim-for-protopia/ Accessed 18 January 2019

Siegfried, T. (2015). Top 10 scientific mysteries for the 21st century. *Science News*, https://www.sciencenews.org/blog/context/top-10-scientific-mysteries-21st-century

Simon, A. (2015). David Birnbaum Cracks the Cosmic Code. *Huffington Post*, https://www.huffingtonpost.com/alex-simon/david-birnbaum-cracks-the_b_7557088.html

Singer, E. (2015). Mongrel Microbe Tests Story of Complex Life. *Quanta Magazine*, https://www.quantamagazine.org/lokiarchaeota-and-the-origin-of-complex-life-20151029 Accessed 5 December 2018

Smith, M.L., Hostetler, C.M., Heinricher, M.M., and Ryabinin, A.E. (2016). Social transfer of pain in mice. *Science Advances*, https://doi.org/10.1126/sciadv.1600855

Softky, W. (1994). Sub-millisecond coincidence detection in active dendritic trees. *Neuroscience*, https://doi.org/10.1016/0306-4522(94)90154-6

Steinberg, G., Schuster, M., Hacker, C., Kilaru, S. and Correia, A. (2017). ATP prevents Woronin bodies from sealing septal pores in unwounded cells of the fungus Zymoseptoria tritici. *Cellular Microbiology,* https://doi.org/10.1111/cmi.12764

Stokes, M.G. (2015). 'Activity-silent' working memory in prefrontal cortex: a dynamic coding framework. *Trends in Cognitive Sciences,* https://doi.org/10.1016/j.tics.2015.05.004

Sul,S., Tobler,P.N., Hein,G., Leiberg,S., Jung,D., Fehr,E. and Kim,H. (2015). Spatial gradient in value representation along the medial prefrontal cortex reflects individual differences in prosociality. *PNAS,* https://doi.org/10.1073/pnas.1423895112

Tagaki, S. (2003). Actin-based photo-orientation movement of chloroplasts in plant cells. *Journal of Experimental Biology,* https://doi.org/10.1242/jeb.00215

Tan, T.H., Malik-Garbi, M., Abu-Shah, E., Li, J., Sharma,A., MacKintosh, F.C., Keren, K., Schmidt, C.F., and Nikta Fakhri1,(2018). Self-organized stress patterns drive state transitions in actin cortices. *Science Advances,* https://doi.org/10.1126/sciadv.aar2847

Tapia-Torres, Y. and Olmedo-Álvarez, G. (2018). Life on Phosphite: A Metagenomics Tale. *Trends in Microbiology,* https://doi.org/10.1016/j.tim.2018.01.002

Temming, M. (2019). People can sense Earth's magnetic field, brain waves suggest. *Science News,* https://www.sciencenews.org/article/people-can-sense-earth-magnetic-field-brain-waves-suggest? Accessed 1 April 2019

Tero, A., Takagi, S., Saigusa,T., Ito, K., Bebber, D.P., Fricker, M.D. Yumiki, K., Kobayashi, R. and Nakagaki,T. (2010). Rules for Biologically Inspired Adaptive Network Design. *Science,* https://doi.org/10.1126/science.1177894

The Royal Society (2016). Summer Science Exhibition 2016: Quantum secrets of photosynthesis. *YouTube,* https://www.youtube.com/watch?v=vBpsHAxsxAg, Accessed 5 December 2018

The Royal Society (2017). Wiring up the brain: How axons navigate. https://royalsociety.org/science-events-and-lectures/2017/03/ferrier-lecture/ Accessed 13 December 2018

Tingley, D., Alexander, A.S., Quinn, L.K., Chiba, A.A. and Nitz, D. (2018). Multiplexed oscillations and phase rate coding in the basal forebrain. *Science Advances,* https://doi.org/10.1126/sciadv.aar3230

Tononi, G. and Cirelli, C. (2003). Sleep and synaptic homeostasis: a theory. *Brain Research Bulletin,* https://doi.org/10.1016/j.brainresbull.2003.09.004

Tsong, T.Y. (1989). Deciphering the language of cells. *Trends in Biochemical Sciences,* https://doi.org/10.1016/0968-0004(89)90127-8

Turchin, P. (2017). Social instability lies ahead, researcher says. *PhysOrg,* https://phys.org/news/2017-01-social-instability-lies.html Accessed 13 March 2019

Tuszyński, J.A., Portet, S., Dixon, J.M., Luxford, C. and Cantiello, H.F. (2004). Ionic Wave Propagation along Actin Filaments. *Biophysical Journal,* https://doi.org/10.1016/S0006-3495(04)74255-1

Tuszyński, J.A. (2014). The need for a physical basis of cognitive process: Comment on "Consciousness in the universe. A review of the 'Orch OR' theory" by Hameroff and Penrose. *Physics of Life Reviews,* https://doi.org/10.1016/j.plrev.2013.10.009

Tyler,S.E.B. (2017). Nature's Electric Potential: A Systematic Review of the Role of Bioelectricity in Wound Healing and Regenerative Processes in Animals, Humans, and Plants. *Frontiers in Physiology,* https://doi.org/10.3389/fphys.2017.00627

University of Waterloo (2014). Weird 'magic' ingredient for quantum computing: Contextuality. *ScienceDaily,* https://www.sciencedaily.com/releases/2014/06/140611131858.htm

Uys,E. (2016). Project Florence: Opening communication with plant life. *Design Indaba,* https://www.designindaba.com/videos/creative-work/project-florence-opening-communication-plant-life. Accessed 12 December 2018.

Uzer, G., Fuchs, R.K., Rubin, J. and Thompson,W.R. (2016). Concise Review: Plasma and Nuclear Membranes Convey Mechanical Information to Regulate Mesenchymal Stem Cell Lineage. *Stem Cells,* https://doi.org/10.1002/stem.2342

Wager, T.D., Kang, J., Johnson, T.D., Nichols, T.E., Satpute,. B. and Barrett, L.F. (2015). A Bayesian Model of Category-Specific Emotional Brain Responses. *PLOS Computational Biology,* https://doi.org/10.1371/journal.pcbi.1004066

Walia, A. (2016). Science Is Finally Proving The Existence Of Meridian Points Throughout The Human Body. *Collective Evolution,* https://www.collective-evolution.com/2016/05/10/science-is-finally-proving-the-existence-of-meridian-points-throughout-the-human-body/ Accessed 1 April 2019

Walia, A. (2019). 5G Is The "Stupidest Idea In The History of The World"- Washington State Biochemistry/Medical Science Prof. *Collective Evolution,* https://www.collective-evolution.com/2019/02/19/5g-is-the-stupidest-idea-in-the-history-of-the-world-washington-state-biochemistrymedical-science-prof/ Accessed 12 March 2019

Walton, M.E. and Bouret,S. (2018). What Is the Relationship between Dopamine and Effort? *Trends in Neurosciences,* https://doi.org/10.1016/j.tins.2018.10.001

Wang, W., Pedretti, G., Milo, V., Carboni, R., Calderoni, A., Ramaswamy, N., Spinelli, A.S. and Ielmini, D. (2018). Learning of spatiotemporal patterns in a spiking neural network with resistive switching synapses. *Science Advances,* https://doi.org/10.1126/sciadv.aat4752

Watson, R.A. and Szathmáry, E. (2016). How Can Evolution Learn? *Trends in Ecology & Evolution,* https://doi.org/10.1016/j.tree.2015.11.009

Weaver, J. (2018). A Unified Theory of Everything Wrong with the Internet. *Medium,* https://medium.com/s/story/the-anonymity-paradox-a-unified-theory-for-what-is-wrong-with-the-internet-673cf6706140. Accessed 12 December 2018.

Wei, S., Evenson, Z., Stolpe, M., Lucas, P. and Austen Angell, C.(2018). Breakdown of the Stokes-Einstein relation above the melting temperature in a liquid phase-change material. *Science Advances,* https://doi.org/10.1126/sciadv.aat8632

Weiner, A. and Eninger, J. (2018). The Pathogen–Host Interface in Three Dimensions: Correlative FIB/SEM Applications. *Trends in Microbiology,* https://doi.org/10.1016/j.tim.2018.11.011

Wessel,J.R. (2018). Surprise: A More Realistic Framework for Studying Action Stopping? *Trends in Cognitive Sciences,* https://doi.org/10.1016/j.tics.2018.06.005

West, G. (2014). Scaling: The surprising mathematics of life and civilization. *Medium,* https://medium.com/sfi-30-foundations-frontiers/scaling-the-surprising-mathematics-of-life-and-civilization-49ee18640a8 Accessed 12 December 2018

Wickstead, B. and Gull, K. (2011). The evolution of the cytoskeleton. *Journal of Cell Biology,* https://doi.org/10.1083/jcb.201102065

Williams, C. (2013). Consciousness: the unconscious is our silent partner. *New Scientist,* https://doi.org/10.1016/S0262-4079(13)61257-9

Williams, C. (2017). Mind the gaps: The holes in your brain that make you smart. *New Scientist,* https://www.newscientist.com/article/mg23331180-300-mind-the-gaps-the-holes-in-your-brain-that-make-you-smart/ Holes in the head, https://doi.org/10.1016/S0262-4079(17)30576-6

Wilson, D. S. and Gowdy, J. (2013). Evolution as a general theoretical framework for economics and public policy. *Journal of Economic Behavior & Organization.* https://doi.org/10.1016/j.jebo.2012.12.008.

Wilson, D.S., Hayes, S.C. Biglan,.A. and Embry, D.D. (2014). Evolving the future: Toward a science of intentional change. *Behavioural and Brain Sciences,* https://doi.org/10.1017/S0140525X13001593

Winawer, J., Witthoft, N., Frank, M.C., Wu, L., Wade, A.R. and Boroditsky, L. (2007). Russian blues reveal effects of language on color discrimination. *PNAS,* https://doi.org/10.1073/pnas.0701644104

Wolpert, J. (2018). The Value of Being Stupid about Blockchain. *ConsenSys Media,* https://media.consensys.net/the-value-of-being-stupid-about-blockchain-c46ba3c99cd6. Accessed 13 December 2018.

Woodward, A. (2017). Your eardrums move in sync with your eyes but we don't know why. *New Scientist,* https://www.newscientist.com/article/2141467-your-eardrums-move-in-sync-with-your-eyes-but-we-dont-know-why/

World Economic Forum (2018). Creating a Shared Future in a Fractured World. *World Economic Forum,* https://www.weforum.org/events/world-economic-forum-annual-meeting-2018/sessions/creating-a-shared-future-in-a-fractured-world Accessed 22 January 2019

Wu, M., Wu, X. and De Camilli, P. (2012). Calcium oscillations-coupled conversion of actin travelling waves to standing oscillations. *PNAS,* https://doi.org/10.1073/pnas.1221538110

Xue, B. and Robinson, R.C. (2013). Guardians of the actin monomer. *European Journal of Cell Biology,* https://doi.org/10.1016/j.ejcb.2013.10.012

Yokawa, K., Kagenishi, T., Pavlovič, A., Gall, S., Weiland, M., Mancuso, S. and Baluška, F. (2017). Anaesthetics stop diverse plant organ movements, affect endocytic vesicle recycling and ROS homeostasis, and block action potentials in Venus flytraps. *Annals of Botany,* https://doi.org/10.1093/aob/mcx155

Zhang, Y., Luo, Y., Zhang, Y., Yu, Y-J., Kuang, Y-M., Zhang, L., Meng, Q-S., Luo, Y., Yang, J-L., Dong, Z-C. and Hou, J.G. (2016). Visualizing coherent intermolecular dipole–dipole coupling in real space. *Nature,* https://doi.org/10.1038/nature17428

Zujur, D., Kanke, K., Lichtler, A.C., Hojo, H., Chung, U-I, and Ohba, S. (2017). Three-dimensional system enabling the maintenance and directed differentiation of pluripotent stem cells under defined conditions. *Science Advances,* https://doi.org/10.1126/sciadv.1602875

Printed and bound by CPI Group (UK) Ltd, Croydon, CR0 4YY